S0-BEZ-062

The Brickworker's Bible

Other TAB books by the author:

No. 872   *Wood Heating Handbook*
No. 892   *Do-It-Yourselfer's Guide to Chainsaw Use & Repair*
No. 949   *Do-It-Yourselfer's Guide to Auto Body Repair & Painting*
No. 1074   *How to Build Your Own Vacation Home*
No. 1144   *Working with Plywood, including Indoor/Outdoor Projects*

Acknowledgement

My special thanks to the Brick Institute of America for providing technical assistance and for the use of many illustrations and tables.

No. 1204
$15.95

The Brickworker's Bible

by Charles R. Self

 TAB BOOKS Inc.
BLUE RIDGE SUMMIT, PA. 17214

FIRST EDITION

FIRST PRINTING—JUNE 1980

Copyright © 1980 by TAB BOOKS Inc.

Printed in the United States of America

Reproduction or publication of the content in any manner, without express permission of the publisher, is prohibited. No liability is assumed with respect to the use of the information herein.

Library of Congress Cataloging in Publication Data

Self, Charles R.
    The brickworker's bible.

    "TAB 1204."
    Includes index.
    1. Bricklaying. 2. Bricks. I. Title.
TH5501.S39   693'.21   80-14452
ISBN 0-8306-9942-2
ISBN 0-8306-1204-1 (pbk.)

Cover courtesy of Brick Institute of America.

Contents

Introduction

The durability of brick construction is almost legendary. During the digging season of 1923-24, Dr. C.L. Wooley of the British University Museum directed excavations at Ur. The excavations revealed the Ziggurat, a shrine to the Moon god, which had been destroyed in about the fifth century before Christ. While clearing the lowest stages of the building of Ur-Engar, Dr. Wooley found a structure, erected about 2300 B.C., rectangular in shape and about 130 by 195 feet in size. Crude brick lined the inside and a fine burned brick lined the exterior. The exterior brick had been laid with a bitumen mortar. Pitch soaked reed mats were used at regular intervals to provide a bond.

According to Dr. Wooley, "The quality of these facing bricks and the bricklaying is astonishingly good. Much of the wall face is as clean and new looking as when it was first built." Such a statement says two major things about brick construction. First, it is one of the most durable of all building products. Second, the process is amazingly ancient. The bricks used at Ur undeniably present evidence of a long period of development.

Findings indicate that brick was actually in use well before 4000 B.C. Probably the first use was by some tool using early man who happened to notice the way sun dried clay formed into irregular cracked blocks. A quick shaping would make those adobe blocks a useful building material.

Over time, methods of producing fire dried, or burned, brick were developed. The basic methods remain pretty much the same today. Clay is first mined, then it is crushed and mixed with water to form a cohesive, thick slurry. Then the clay is molded into the proper sizes, air dried for a time and then fired. Naturally, methods of production have changed with the years, but the essential process remains the same today. Modern mechanization has brought change that now allows the production of billions of bricks a year. In 1979, over 9 billion bricks were manufactured in the United States.

While it might seem true that "a brick, is a brick" and so on, a quick look at the sample bricks at your local building supply outlet will quickly change your mind on that score. The variety of designs, sizes, weights and uses is sobering. It is now possible to build almost anything with brick, especially in combination with other materials.

The uses for brick are nearly unlimited, whether simply for decoration or for structural strength. Modern burned brick comes in almost 10,000 different styles and colors, with a wide, wide variety of shapes and colors. It can be used to construct a child's sandbox, a patio, a retaining wall, a solid house wall, a brick veneer house wall, interior floors or steps. About the only place it might not prove too handy is on a ceiling—though even there some of the new "thin" bricks, about an eighth the thickness of standard brick, could be used.

Its beauty is an excellent reason for choosing brick over other materials or in conjunction with other materials. Another advantage is the fireproofing quality of brick. It just won't burn in any normal fire since the material has already been heated to 1900 degrees Fahrenheit. Using brick to build a home will usually result in lower fire insurance than will standard frame construction techniques.

Cost often seems a factor in rejecting brick, but a large portion of the cost of brickwork on or around a house is the price of labor. A good brick mason seldom makes much less than $10 an hour anywhere in the country, and in many areas he or she will make much more.

While many people view the use of brick as a field only for professionals with extensive training and experience, almost anyone with normal capabilities and a fair amount of care can learn to use this ancient building material. Most people *can* do brick masonry, but be forewarned that it is not light work by any means.

Charles R. Self

Brick Making

The production of brick today is quite different than it was in the days when workers pressed their fingers in the still damp material as a signature. Plano-convex marked bricks date from some 4000 years B.C. and special signature stamps began to come into use on the bricks of Ur around 2250 B.C. In most early cases, the making of brick was declared a gift of a particular god, and the entire process of manufacturing was often taken under the patronage of the particular royal monarch of the time.

The ancient art of brick making is thought to have spread from the Mesopotamian plains into Persia, India, China and then west to Egypt, Greece and Rome. Biblical references to building with brick abound. Herodotus, the Greek historian of the fifth century B.C., provides a detailed account of the wonders of the Babylonian walls and hanging gardens which Nebuchadnezzar had built with brick. The walls were immense. The inside section was about 23 feet thick and made of sun dried brick. There was a mud filled section over 39 feet thick and a facing wall of burned brick about 25 ½ feet thick. The entire mass, and a mass it was, stood over 365 feet tall!

Europe received the art of brick making from the Romans during their three and a half century occupation of the continent and England. In England, the practice of brick making just about disappeared until the 13th century. It was not until the reign of Henry VIII that brick making really began to flourish once again. When London burned in 1666, the brick industry gained immensely. What had been a city built primarily of wood became a city built primarily of brick. By the 18th century, the Georgian influence with its fine brick country houses virtually shut out any other form of building material.

The first brick in America were the adobes the Spanish Conquistadores found in Peru and Mexico. In Virginia, the brick industry started about 1611. In Massachusetts a brick making industry began about 1629.

All in all, it is easy to see why brick is known as the oldest manufactured building product. While methods have been mechanized and speeded up, there is essentially little difference in the basic work than back when the brickmakers dug the clay by hand and used wooden molds to form the bricks. Up to a point. Much of the early brick used was adobe, or sun dried, while virtually all modern brick is fired or burned brick that is run through vast oasts, or kilns. Originally, brick was fired by building a small hut from the green brick and then building a fire inside the hut. Real change didn't begin to come until the Industrial Revolution of the last century. Actually, the first brick making machine recorded pre-dated the 19th century by a few years. Apollos Kinsley, of Connecticut, patented a machine that was to influence brick making for at least 50 years beyond its patent date of 1793. Essentially, Kinsley's machine was a vertical mill with an augur to force the clay into molds underneath. The molds were larger than forced clay column and had a small hole in the top to allow the sprue to show when the mold was filled. By 1798, an Englishman named Farquharson had developed, and patented, a slightly more sophisticated version of Kinsley's machine that could be driven by a horse. Around Washington, D.C., in 1819, a horse driven dry press with a circular pressing table, was in operation. This press was said to be capable of producing brick at the rate of 30,000 per 12 hour day, using a single horse. The number of men required to hand fill the molds is not available.

Dry presses were important simply because in those days most brick, when wet pressed, was not ready for the kilns until some of the excess water was air dried. Since the brick was often dried in open sheds, much production could be lost to inclement weather.

In the 30 years before 1820, the U.S. Patent Office received 21 brick making machine patents. In the next 16 years, there were 69 such patents.

By 1862, the extrusion process and wire cutter had been invented. The basic process continues with little change today.

Naturally, machine size, power, more efficient lubrication and other factors have improved so that extruded brick is the rule today.

Kilns were often a problem during early days of brick making. Heat conservation in the early models was nearly impossible and this raised the cost of brick manufacture quite a lot. If those early inefficient kilns were in use with today's very costly energy, few except the wealthiest people, would be able to afford brick as a building material. Continuous kilns, or tunnel kilns, were first built around 1751. The early kilns had such a great capacity the molds couldn't keep them in use. This cut the advantages of heat conservation to nil since the kilns had to be shut down frequently and then re-fired.

A circular tunnel kiln was patented by a Swiss engineer in 1857. A straight line tunnel kiln was built in 1866. It wasn't until about 1915 that tunnel kilns became really practical. It was at this time that tools were developed that allowed for the optimum use of these kilns.

As it can be seen, almost all of the basic tools for today's brickmaking were introduced before the turn of the century. Development has continued and efficiency has increased. Today's brick, in relation to inflation, is probably cheaper than at almost any other time in history.

At one time there were 54 brick factories in a single town, Haverstraw, on New York's Hudson River. There are now probably no more than a couple hundred across the entire country. Incidentally, the brick factories in Haverstraw at one time took so much clay from the river bank that a large portion of the town slid into the river, killing 10 people in the process.

The brick manufacturing process requires clay with enough plasticity to allow shaping or molding when mixed with water. Also, the clay must have sufficient tensile strength to hold a shape after forming is completed.

TYPES OF CLAY

There are three primary forms of clay. *Surface Clays* are probably upthrusts of older deposits and they are found, as their name indicates, near the surface of the earth. In south central Virginia, I'm almost surrounded by good ol' red surface clay, which, oddly enough, is also quite fertile if handled correctly. *Shales* have been subjected to high pressures so that the clay is hardened almost to the consistency of slate. *Fire clays*, which are found at deeper levels, must be mined. Fire clays are important to fireplace builders and wood stove manufacturers since they have refractory properties (fire and heat resistance) which keep them from cracking under extreme heat conditions. Generally, there are fewer impurities in fire clays. Essentially, all clays have similar chemical compositions even though the physical form can differ markedly. All are compounds of silica and alumina, with varied amounts of metallic oxides and other impurities. The impurities, oddly enough, actually act as fluxes allowing fusion at much lower temperatures. One of the reasons fire clays ward off the effects of extreme heat so well is the lack of impurities. Fusion occurs at much higher temperatures than for standard brick and this creates the needed qualities in fire brick. Iron, magnesium and calcium have the strongest affect (at least their oxides do) on the color of the finished brick.

MANUFACTURING BRICK

The manufacturing process for brick has six stages:
—Winning and storage of raw material.
—Preparation of raw materials.
—Forming units.
—Drying.
—Burning and cooling.
—Drawing and storing the finished units.

Winning is simply the term applied to mining procedures in the clay industry. Common practice is to hold several days worth of raw materials in storage so that poor weather conditions can't force an expensive shutdown of the equipment.

Preparation involves crushing the clay to break up large chunks, removal of stones and then moving the clay to huge grinding wheels. The grinding wheels weigh from four to eight tons. The material is ground and mixed and the clay is usually screened in order to control particle sizes and get better uniformity in the final bricks.

Forming starts with tempering to produce a homogeneous plastic mass. This is done by adding water to the clay in a mill with a mixing chamber having one or two blades. There are three processes used for forming. They are stiff mud, soft mud and dry press. The stiff mud process clay is mixed with just enough water to produce plasticity—usually twelve to fifteen percent by weight. Once the clay and water are thoroughly mixed (pugged), the clay goes through a de-airing machine with a vacuum of 15 to 29 inches of mercury held. This results in fewer air holes, increased workability and greater final strength. The clay column is forced through a die and then passes through an automatic cutter. Cutter wire spacings and die sizes are carefully set up to make up for normal shrinkage from the wet stage through the final burning. After leaving the cutting table, the units are inspected. Those with imperfections go back to the pugging mill, while the perfect units are placed on a drying car.

The soft mud process is used only for the production of brick. The stiff mud process can also be used to produce structural clay tile. The soft mud process is most often used when clays contain too much natural water for the stiff mud process. Clay is mixed with 20 to 30 percent water, with the units then formed in molds lubricated with sand or water. This is the oldest form of producing brick.

The dry press process is used where clays have very low plasticity, using a minimum of water (no more than 10 percent). The units are then formed in steel molds under pressures ranging from 500 to 1550 pounds per square inch.

When wet clay bricks come from the molding or cutting machines the water content, by weight, will range from seven to 30 percent, depending on which of the three above forming methods has been used. Before *burning* can begin most of this water is evaporated in drier kilns, using temperatures from about 100 degrees to 400 degrees F. Drying time will vary from 24 to 48 hours. The heat for the drier kilns is most often tapped from the waste heat produced by the burning kilns. Heat and humidity is carefully regulated to prevent cracking.

A few bricks, and most tile and terra cotta, are glazed. High fired glazes are sprayed on the units either before or after drying and the units are then burned at normal firing temperatures. Low fired glazes are used where colors would be lost in extreme heat. They are applied after the unit has been burned and cooled. The units are then re-fired to relatively low temperatures.

Burning requires from 40 to 150 hours of the clay unit production time, depending on a number of variables. Dried units are placed in the kiln in a prescribed pattern to permit free circulation of hot air. In tunnel kilns, units are loaded in special cars in a similar manner. Burning divides into six separate stages:

— Water smoking (evaporation of free water).
— Dehydration.
— Oxidation.
— Vitrification.
— Flashing.
— Cooling.

All except flashing and cooling require rising temperatures in the kiln. Water smoking takes place at up to about 400 degrees F. Dehydration takes place at about 300 to 1800 degrees F. Oxidation takes place at from 1000 to 1800 degrees F. Vitrification takes place at from 1600 to 2400 degrees F.

The rate of temperature change must be carefully controlled. The proper temperature depends on the raw material and the type of unit being produced. Kilns are equipped with recording pyrometers to keep a careful check on the burning process.

Near the end of the burning process, the units can be flashed to change the color. In essence, flashing means that clays, regardless of their natural color, containing iron oxides which will burn to a red color when exposed to an oxidizing fire as ferrous oxide forms. Changing the fire to a reducing type will give the units a purplish tinge. It is the use of a reducing fire that is known as flashing.

Cooling is another step that must be carefully controlled. Tunnel kilns will usually allow a finished cooling in about 48 hours or less. Care must be used since a too rapid cool down will cause excessive cracking. The next step after cooling is *drawing*. This is the process of removing the bricks from the kiln. Then the bricks are sorted, graded and either stored or shipped.

CHARACTERISTICS OF BRICK

Both the raw materials used and the manufacturing process affect the final properties of brick and clay tile. The properties to be most concerned with in these materials are color, texture, size variation, absorption, compressive strength and durability. For firebrick, the refractory qualities are also important.

The color of burned clay depends on its chemical composition, the burning temperatures and the type of burning control used. Iron has the most effect on color since a normal, or oxidizing, fire will always produce a red brick when iron is present. The higher the temperature of the burning kiln, the darker the color and the lower the absorption. In addition, compressive strength is increased.

Texture can be obtained by using dies or molds during forming (Fig. 1-1). Smooth brick is usually called die skin and it is a result of the pressure exerted in the steel die (Fig. 1-2). If the stiff mud process is used to form the brick, attachments can be used to add cuts, scratches, brush marks or other textures to the face of the brick as the clay column leaves the die.

Sizing is important with brick since you would naturally like to know, within reasonable limits, just how many bricks are going to be

Fig. 1-1. Wall design variations are almost infinite with brick.

needed to do a particular job. Different clays shrink at different rates under different conditions. Air shrinkage will range from 2 to 8 percent, while fire shrinkage can move from 2.5 to 10 percent. Total shrinkage might cover a range from 4.5 to 15 percent. Fire shrinkage is greatest at higher temperatures.

Compressive strength and absorption are qualities affected by the clay properties, method of manufacture and degree of burning. Generally, the stiff mud process will provide units of greater compressive strength and lower absorptions than units produced by the other two processes. However, there can be exceptions. Given a particular type of clay, the higher the burning temperature the greater the compressive strength and the lower the absorption. The type of clay used has a great amount of influence on the final outcome (Fig. 1-3).

So far, I have pretty well covered the manufacture of the most popular types of brick for home and commerical use in this country. But another form seems to be regaining its former popularity among some groups devoted to "natural" living. Air dried, or adobe brick, has often been used for construction by Indians of South America and the American Southwest. Adobe brick certainly requires less energy, in the form of heat, to produce.

ADOBE BRICK

Essentially, adobe brick is a mixture of claylike mud, finely chopped straw, and water that is left to dry in the sun. It is, therefore, more suitable for production and use in those areas of the country, such as the Southwest, where sunny days tend to prevail. To my knowledge, at present no one is producing such brick on a large scale, so you'll almost certainly need to do your own adobe brick manufacturing. Molds are used and the actual mixture has to be a matter of experimentation. The consistency your clay will determine much of the other requirements. The batch is mixed very well, poured into the molds and allowed to dry in the sun until completely hard.

Adobe bricks should be laid using the same mortar you would use for burned brick or concrete block. The same basic techniques are used—with one note. Generally, adobe bricks are usually larger and heavier than standard brick. Some weigh as much as 40 or more pounds per brick and a few weigh under 20 pounds. Also, adobe is noted for its ability to absorb water. This can, in heavy rain areas, quickly reduce a building to little more than a muddy mass on your lot. A treatment of some kind is almost always needed to provide water resistance if the area is subject to heavy rains or flooding.

Fig. 1-2. Exterior brick uses are extremely varied and are by no means limited to simple walls.

Fig. 1-3. As an accent material brick has no peer.

A lot of estimates can be found on the supposedly unparalleled insulating qualities of brick—expecially adobe brick. Don't believe them. No solid material of the type of construction used for brick, whether adobe or burned, offers very great insulating properties. Generally, a 2- to 3-foot thick brick or adobe wall would be needed to equal the insulating properties of a house framed in 2 × 4s, using 3-inch thick fiberglass insulation between studs.

Brick Styles & Sizes

Much of the success of a particular bricklaying job will depend on the selection of the proper brick size and the correct style to fit both the decor of the area, or home, and the job at hand (Fig. 2-1). It would be a virtual impossibility to cover each and every brick style, not to mention special sizes on special orders from smaller brickyards. However, there are some constants in this field and a normal brick size—though the actual size of the brick is somewhat less than the standard module.

Standardization in the sizes of building materials is now something of a generally accepted rule, though far from perfect. Since the construction industry in most cases still operates on the English system, without bothering with metrics, you can, for the most part, do the same. It is possible to have your supply house or hardware store order metric, or metric/English measuring equipment from tool makers such as The Stanley Works. But right now, and in the immediate future, what you have should serve.

In most cases, modules run in units of four, either 4 inches or 4 feet. Examples are 16-inch on center framing, 2-foot on center framing and 4 × 8 foot panelling. In the case of bricks, the modules are 4 inches, though the actual brick is smaller. If it weren't, the mortar would make the module oversized. Standard bricks are nominally sized at 4 inches by 4 inches by 8 inches, while their actual size is 3 ⅝ inches by 3 ⅝ inches by 7 ⅝ inches. With such sizing, the brick manufacturers have allowed you to place a three-eighth inch thickness of mortar on three sides of the brick.

Fig. 2-1. Exterior brick use and design is varied. Several styles of brick can easily be used in the same area.

It should be remembered that this is a basic brick and there will be variations from maker to maker. There will also be variations within a lot of bricks, since materials will vary, as will burning time and other things. Also, the likelihood of variation in size from one maker to another is great. It usually pays to buy all the brick for a particular job from the same manufacturer. It is best if all the brick is bought in a single lot, as this method comes closest to assuring you the greatest consistency.

In the United States, the standard brick is 2 ¼ × 3 ¾ × 8 inches in size, while English bricks are 3 × 4 ½ × 9 inches and Roman bricks are 1 ½ × 4 × 12 inches. Norman bricks are 2 ¾ × 4 × 12 inches. These are pretty much the standard brick styles and sizes, but there are other types of building brick classifications. One of those classifications is *building brick.*

Building brick is what was formerly known as common brick. It is made of standard clay or shale, with no special markings and without special color or surface texture. This type of brick is commonly used for the backing courses in solid and cavity brick walls.

For exposed surfaces, *face brick* is most often used. Face bricks are of basically higher quality than building bricks, with better durability and better looks. Such brick is usually readily available in a variety of shades of red, brown, gray, yellow and white. And many patterns and textures.

For rough work where great durability is important, you can ask for *clinker brick.* Clinker brick results from over burning in the kilns. The bricks are often not very regularly shaped and this class is sometimes called rough-hard.

Pressed brick is made with the dry press process. It has regular, smooth faces, sharp edges and perfectly square corners. It is almost always used as face brick.

For fancier applications, *glazed brick* can be purchased in a variety of colors. One surface is given a ceramic coating of mineral ingredients which fuse together to produce a glass-like coating when the brick is burned. In most cases, such bricks are used where a smooth, easily cleaned and attractive surface is desired. The use of the type of brick will often save the cost of covering a wall with ceramic tile.

Fire brick is used for lining fireplaces, woodstoves and barbecues. Although most people don't bother to line barbecues, their durability can be greatly increased if fire brick is used. As I mentioned in the first chapter, this type of brick, made from deep mined fire clay of greater purity than other clays, is strongly resistant to the effects of heat. It requires much greater heat to get fusion, which makes the brick harder and less likely to crack and decompose under regular heating and cooling cycles. Fire brick is almost always larger than standard brick, with a nominal size of variance depending on the maker. Most fireplaces installed today have metal liners instead of firebrick. This reduces the need for a lot of work, and adds the possibility of increasing fireplace efficiency. Today, no one at all serious about wood heat will even consider using an unmodified fireplace. Even most of those that are modified are nowhere near as efficient as a good, modern wood stove.

Most types of brick can also be obtained in what is known as the *cored brick* style. Basically, cored brick has either three holes in a single row up the center or two rows of five holes. This does little more than cut the weight of the bricks since there is no appreciable difference in strength or resistance to moisture penetration between cored and uncored bricks. Use whatever type is most readily available in most suitable style, color and texture for the job at hand.

Occasionally you might run into some types of imported brick, *European brick*, especially those from England and Holland, that compare favorably with American clay brick as to strength and durability. The price might be somewhat excessive unless only a few are needed. Even more occasionally, you might turn up some *sand-lime brick*. These bricks, most extensively used in Germany, are made from a thin mixture of slaked lime and very fine silicious sand molded under pressure and hardened by steam.

STRENGTH

A great deal of the strength in any brick laying job depends on the mortar used, but there are also other important factors. The strength of the brick is very important. Well burned bricks often exceed a compressive strength of 15,000 pounds per square inch. The workmanship of the bricklayer has much to do with the final job strength. For that reason, there should be much emphasis on using care in every aspect of the bricklaying job. This includes starting with the first, basic layout of the job through mixing the mortar and laying the brick.

If the brick is not uniform, the strength of the job will be decreased. It is a good idea for you to keep an eye on the load as it is delivered and as you go through the pile. Order enough extra bricks to make sure you can set aside the ones seriously out of uniformity with the rest of the batch.

The method used in laying the brick—here I mean the type of bond used, not necessarily the workmanship—has a great deal of effect on job strength. Bonding is covered more thoroughly a bit later. It has at least three different meanings when applied to brick masonry, but essentially here I am using the term to mean the pattern in which the bricks are laid, one upon the other.

Table 2-1. Fire Resistance.

Normal wall thickness (inches)	Type of wall	Material	Ultimate fire-resistance period. Incombustible members framed into wall or not framed in members		
			No plaster (hours)	Plaster on one side* (hours)	Plaster on two sides* (hours)
4	Solid	Clay or shale	1¼	1¾	2½
8	Solid	Clay or shale	5	6	7
12	Solid	Clay or shale	10	10	12
8	Hollow rowlock	Clay or shale	2½	3	4
12	Hollow rowlock	Clay or shale	5	6	7
9 to 10	Cavity	Clay or shale	5	6	7
4	Solid	Sand-lime	1¾	2½	3
8	Solid	Sand-lime	7	8	9
12	Solid	Sand-lime	10	10	12

*Not less than ½ inch of 1:3 sanded gypsum plaster is required to develop these ratings.

Certain types of brick are fairly porous and are known as high suction types. If these bricks are not wetted down before being laid, they will suck moisture into the brick from the mortar more rapidly than the mortar can cure and form the mortar bond. This would seriously weaken the structure.

FIRE RESISTANCE

Tests have been conducted on brick walls with portland-lime mortar (recommended for greatest strength in almost every job). The tests show great fire resistance under the normal testing standards of the American Society for Testing Materials. With only a singe brick 4 inches thick with no plaster, wall fire resistance is an hour and a quarter. Table 2-1 shows the resistance of other wall thickness, with and without plaster on one or two sides of the wall.

INSULATING CHARACTERISTICS

The heat insulating properties of brick are not great. Insulation normally depends on providing a layer, or layers, of unmoving air. Once the air molecules move, they can transmit heat. Brick cannot do this job. If a solid, or unframed, brick wall is to be used, it should have a cavity in which insulation can be inserted. Otherwise, the brick wall should be backed by a framed, insulated wall. This is known as *brick veneering.*

The massiveness of a full brick wall provides fine sound insulation. However, testing shows that increasing the size of a wall more than a foot thick indicates there is no point in building heavier simply to muffle sound. The reduction of sound is not appreciable after the wall is 10 inches thick. The use of cavity walls, as recommended for walls to be insulated, will cut down on sound transmission at lower cost. An added advantage is lower heat loss. While sound transmission through brick walls is low,sound originating inside a brick walled building (surfaced, at least, with brick on the interior) will rebound quite well. Sound deadening is not a quality of brick. Impact sounds, which are sounds resulting from an object striking the wall, travel quite far along a brick wall.

As with every construction material, brick masonry will expand and contract with changing temperatures. Therefore, walls over a few hundred feet in length will probably need expansion joints. However, very few homeowners will be working with such long structures.

Abrasion resistance of brick is related to compressive strength, which is increased by proper burning. If brick is to be used as a floor, for example as an outdoor patio, it should be well burned for greatest durability. Solid brick, depending on the clay used and the degree of burning, will weigh anywhere from 100 to 150 pounds per cubic foot. Generally, the heavier the brick, the better burned it is.

3

Basic Bricklaying

A good job of bricklaying depends on many things other than the quality and style of the brick. For the homeowner, too many worries about efficiency can be self defeating. Often it is necessary to not only cart your own bricks to the point of construction, but you must also mix your own mortar, soak the bricks, set any needed ladders or scaffolding and then lay the bricks. The professional brick layer will almost always have a helper to assist him with or to do such jobs.

Still, some steps toward efficiency will help. Keep the portland cement under cover, of course, and keep it as close as possible to your sand and lime piles. Both sand and lime must be well covered against inclement weather. The lime will cake when wetted and dried. Overly wet sand usually means that getting a properly elastic mortar mix is impossible. All material should be as close to the actual construction site as possible.

Most people working with bricks and mortar use some sort of wheelbarrow as a mixing unit. If you want a good job, don't use partially set up mortar. It is best to mix no more mortar than you will need at one time. Fifty bricks will be the amount, or very close to it, that you can do if you are using a single sack of pre-mixed mortar. A source of water nearby is a help, but buckets and hoses will work if no tap is close. Keeping tools and mixing tubs as free of partially dried mortar as possible will also be an aid to a good job. Frequent clean up will add somewhat to the time needed to complete the job.

If you will need scaffolding, you should plan ahead and gather materials before work begins. Often, little more than a few 2 × 6 or

Fig. 3-1. Even a small amount of brickwork surrounding a fireplace can add to the coziness of a room.

2 × 8 boards across some sturdy sawhorses are all that is needed. Be very careful while working on any type of platform or scaffold. It might not seem all that dangerous and it usually isn't, but special care is necessary whenever you are working above ground level.

If brick is stacked along a wall in amounts you anticipate needing, you will save a great deal of running around time as work goes on. Brick weighs over 100 pounds per cubic foot and the mortar is quite heavy. If you use some sort of platform to hold the brick and mortar supply as the wall, chimney or other object rises you can avoid an awful lot of backache that comes from bending. I recently completed an indoor decorative chimney where there was not room enough to work in such a manner. By the time the chimney was seven feet tall, the resulting backache was made bearable only by the overall change in the looks of the room.

BONDS

Most of us tend to view the word bond as a simple thing, meaning simply how well the mortar holds the bricks or other units together. Actually, there are three types of bonds in brick masonry. One of these categories has six subcategories.

First comes the structural bond, the method by which individual masonry units are interlocked so that the entire structure acts as a single assembly. Without this structural bond you have only a wall or other structures that topple at the first hint of high winds or a sharp blow. There are three ways in which structural bonds are accomplished. Overlapping or interlocking the masonry units is one method. Metal ties can be imbedded in connecting joints, for the second type of structural bond. A third method is the adhesion of grout to an adjacent horizontal masonry wall.

The mortar bond is as important to structural integrity as the structural bond and the adhesion of the joint mortar to the bricks tiles, cement block or to any reinforcements such as veneer wall brick ties.

The pattern bond is simply the pattern the bricks and mortar joints form in the face of the wall (Fig. 3-1). It might or might not be related to the structural bond method used. In other words, the pattern bond might be an integral part of the overall structural integrity of the wall, or it might simply be a decorative bond with other methods used to form a good structural bond (Fig. 3-2). Essentially, there are five basic pattern bonds:
—The running bond.
—Common (American) bond.

Fig. 3-2. Turn a corner and create extra nooks for firewood, an indoor barbecue or an oven.

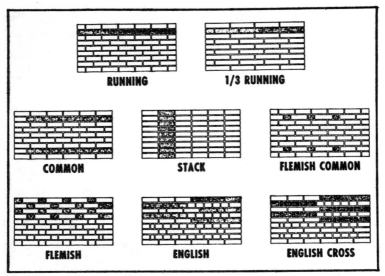

Fig. 3-3. Types of masonry bonds.

—Flemish bond.
—English bond.
—Stack bond.

Figure 3-3 shows several types of bonds in addition to the patterns for the five basics.

Running bonds are the simplest for the layman to use, since all stretcher bricks are used, and each brick runs halfway across the next brick. The running bond is often used with metal ties—for veneer walls. It also is a handy bond for cavity wall construction (Fig. 3-4). A variation is the one-third running bond, usually using a longer than standard brick, with one upper course brick overlapped one-third of the stretcher length. The next course covers the final two-thirds of the lower course brick. Either type of running bond is fairly easy to do. This method requires only minimal advance planning and layout to get everything fairly exact.

Common, or American, bond is something of a variation on the running bond. Instead of all stretchers, a course of headers is added every so often. Usually the course of headers falls at regular intervals, although this can be varied as you wish. For example, a course of headers—where the brick is turned sideways and only the end shows in the course—could be placed on the third course up, another on the eighth course up, another on the eleventh course and so on. The common bond is a bit harder to work out and lay than the running bond alone, but the header courses provide a full structural bond either to brick ties or to the second section of a cavity wall. The structural needs will often determine the header course pattern. Some walls might have to have a header course at every fourth course, while others will not need a header more than every seventh course or so. Starter brick for header bond are known as three-quarters, so even on even runs brick must be cut for this type of bond.

Flemish bond offers both a structural and decorative option—as does common bond which can be done in the same manner. With Flemish bond, headers are used as alternates to stretchers on every course. The headers are centered over the stretchers in the course below. For purely decorative use, bricks can be cut in half and used as headers. For structural use, they can be left whole and tied into either a second wall, for cavity wall construction, or to metal brick ties for veneering. Half brick used in this way are called *blind headers*. There are two different methods of starting the header courses in Flemish bond. Using three-quarters of brick is

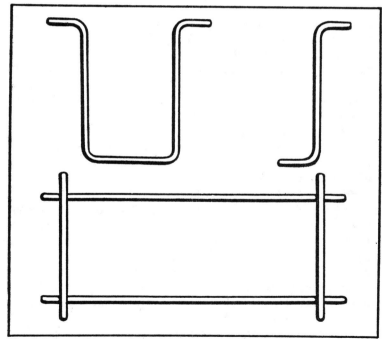

Fig. 3-4. Several types of metal ties used to add structural integrity to brick masonry.

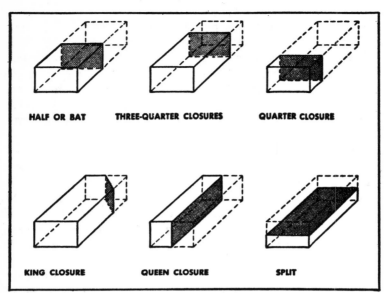

HALF OR BAT   THREE-QUARTER CLOSURES   QUARTER CLOSURE

KING CLOSURE   QUEEN CLOSURE   SPLIT

Fig. 3-5. The shapes of cut brick.

called the Dutch method, while the use of a two-inch (quarter brick) closure is known as the English method.

English bond consists of using alternating courses of headers and stretchers. This differs from American bond in that the headers are centered on the stretchers and the joints between the stretchers line up vertically. Blind headers can be used in any course which has no need for structural bonding.

Block or stack bond is purely and simply a pattern bond. It offers little structural strength, since all vertical joints are aligned. In most cases, rigid steel ties are essential to structural integrity when using stack bond. However, it is often possible to buy and use 8-inch thick stretchers to add the needed strength. Even with the 8-inch stretchers or the steel ties, if the wall is to be a large one and load bearing, steel pencil rods are needed to provide appropriate strength. These pencil rods are inserted in the horizontal mortar joints. For the best appearance with stack bond, you'll need to carefully select the bricks to be used. They must be closely matched to make this geometric pattern attractive. Personally, no matter how well done it is, I find stack bond walls boring because of the extreme regularity.

The English cross or Dutch bond is a mild variation of the straight English bond. It differs only in that the vertical joints between the stretchers in alternate rows don't line up vertically. Instead, the vertical joints center on the stretchers both above and below.

MASONRY TERMS

A further description of some of the non-standard vocabulary used with various types of masonry work might well be in order about now. Few of us, before getting involved in brick masonry, even know what the different faces of a brick are called. Brick masonry is essentially the construction of almost any project using units of baked clay or shale in a uniform size that are laid in courses with mortar joints to form the structural units. Cut brick is named by shape, with the half brick also known as a *bat*. Then there are three-quarter and quarter closures, a king closure (cut at an angle across one corner), the queen closure (cut in half lengthwise) and the split, which, as its name implies, is a half brick split up the middle (Fig. 3-5).

Brick surfaces also have their appropriate names, as Fig. 3-6 shows.

A *course* is nothing more than one of a continuing row of bricks, which, when bonded to other courses, forms the basic masonry structure (Figs. 3-7 and 3-8).

A *wythe* is a continuous vertical, four-inch or wider, section or thickness of masonry. Probably the best example is the wythe surrounding a chimney to keep flues separated.

A *stretcher* is a brick laid flat with the longest part parallel to the face of the structure being built.

A *header* is a masonry unit (brick) laid flat with its longest side perpendicular to the face of the unit under construction. Headers

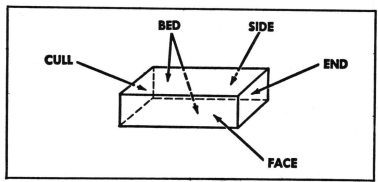

Fig. 3-6. Names of brick surfaces.

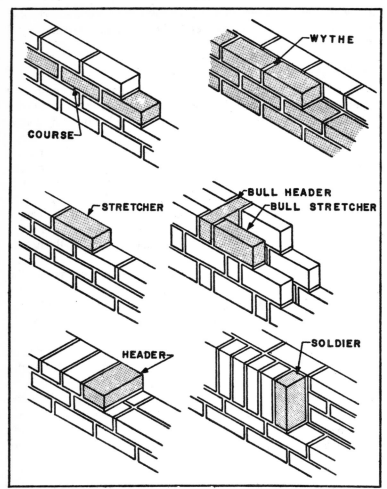

Fig. 3-7. Masonry units and mortar joint names.

are most often used to tie two wythes of masonry together, but extra length headers can also be used to form a cavity wall.

A *rowlock* is a brick laid on its face.

A *bull-stretcher* is a brick laid with its longest side parallel to the face of the wall.

A *bull header* is a rowlock brick laid with its longest side perpendicular to the face of the wall.

A *soldier* is a brick laid on its end so that its longest dimension is parallel to the vertical mortar joints of the wall.

Fig. 3-8. A running bond interior wall (courtesy of Structural Clay Products Institute).

Fig. 3-9. The proper way to hold a trowel.

Flashing in masonry is almost always of metal and usually aluminum. It is used to move moisture away from spots where the masonry is particularly vulnerable to penetration from water. Generally flashing is used under horizontal masonry surfaces such as sills, where masonry walls intersect with the roof and other such surfaces (around the chimney base, for example), over door header openings and window headers, and sometimes along floor lines. Flashing extends through the outer face of the wall and is turned down, with weep holes provided every eighteen to twenty-four inches so that water pooling on the flashing can drain. Generally, no matter the requirements of appearance, it is much more sensible to extend the flashing beyond the face of the wall. Concealed, or partially concealed, flashing could concentrate wetness and this would harm structural integrity.

THE BASIC MORTAR JOINT

A bit later I will go into further details of mixing mortar and the layout of more complex brick structures, but one of the most important techniques to work on with brick masonry is laying and tooling the basic mortar joint. As a start, holding the trowel correctly—when working space permits—is a help. The thumb does not, and should not, circle the trowel grip. The trowel is generally pointed down and away from the body (Fig. 3-9). If you

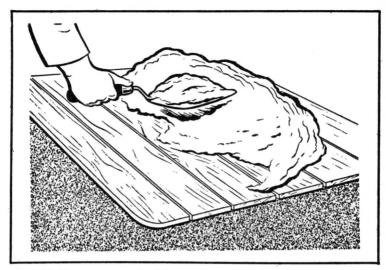

Fig. 3-10. The proper way to pick up mortar.

are right handed, you will then use the left edge of the trowel to pick up your mortar—from the outside of the mortar pile. Pick up enough mortar to cover no more than three bricks at the outset. Even less mortar is better for the novice. Picking up enough for a single brick will form a small row along the left side of the trowel, while enough mortar for five bricks will load up the standard sized trowel (Fig. 3-10).

Fig. 3-11. A trowel full of mortar in position for placement.

Fig. 3-12. Mortar thrown on brick.

The left edge of the trowel is held directly over the center of the course of bricks, tilted and the mortar drops into place as the trowel is moved to the right (Figs. 3-11 and 3-12.) Any mortar remaining on the trowel is tapped back onto a mortar board for later use, not onto the course of bricks. Of course, instead of working from left to right, a left handed bricklayer would simply reverse the above procedures (Fig. 3-13.)

Any mortar hanging over the edges of the brick is scraped off and returned to the mortar board, with the exception of enough to "butter" the end of the next brick to be laid in this course (Fig. 3-14.)

The bed, or base, joint is spread about an inch thick with mortar. You then make a shallow furrow, tapering from the center to the outsides, butter the end of the first brick and push it into the mortar. The furrow must be shallow. If you make it too deep, the gap left between the mortar and the brick will allow for excessive moisture penetration. Eventually, and especially in exterior construction, this will result in the joint breaking up and losing its bond. Mortar for the bed joint should be spread no more than five bricks along the course. Mortar that is spread too far dries too rapidly, cutting back on the bonding properties. If the mortar isn't soft and plastic, scrape it up, clean the bed and re-start your bed joint. Any mortar that has dried this far should not be returned to the mortar board (Figs. 3-15 and 3-16.)

Fig. 3-13. Spreading mortar.

When the brick is picked up to be bedded in the mortar, your thumb will be on one side and your fingers on the other. The brick is held with the end in the air and the end is buttered. Buttering consists of pushing as much mortar on the end of the brick as will stick. If you don't get enough mortar on the end of the brick, your

Fig. 3-14. Excess mortar cut off.

Fig. 3-15: Mortar spread and the bed joint started.

head joint (vertical joint) will not be full. Push the brick hard enough
to force mortar out of the head joint. Then cut off the excess mortar
and return it to the mortar board. If the mortar is setting up a bit, but
not a lot, it can be set on the back of the mortar board for possible

Fig. 3-16. A bed joint furrow.

Fig. 3-17. The proper way to hold a brick for buttering.

later retempering—more on retempering later (Figs.3-17, 3-18 and 3-19).

Of course, a mason's line is used to get the horizontal positioning of the bricks correct. Experienced brick masons will often use

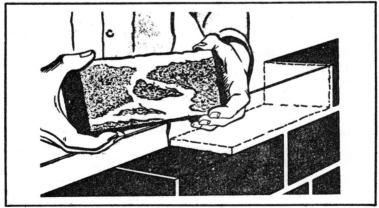
Fig. 3-18. A poorly bonded brick.

Fig. 3-19. Setting the head joint in a stretcher course.

such a guide only every eighth or tenth course of bricks, but for novices it is probably best to run the line again after two courses have been laid. In fact, if you seem to have trouble keeping the course horizontal, move the cord after each course is down. It might seem fussy and a time waster. But in the end, the results will be well worth the extra time taken. Vertical wall plumb can be held with a plumb bob on a cord or with a level used after every few of courses. Actually, the level can be used to hold the horizontal line too, but the line is a bit easier and more rapid to use in most cases. This is expecially true on long courses where you can move along without checking the level of every two or three bricks.

Occasionally, there will be some reason you need to insert a single brick in a space in a wall. To make certain there is a good all around bond, be sure to lay a thick bed of mortar on the bottom of the hole and on the sides (Fig. 3-20). Then lay a thick buttering of mortar on the top of the brick and shove it into place. When the brick is shoved to its final position, you should find mortar squeezed out of each joint, including the top, bottom and both headers. The excess mortar is, as always, then scraped off.

For joining wythes, you'll need to make cross joints in header courses. Particularly if the header course is to aid in structural integrity, you must make certain these cross joints are full of mortar. The entire side of the brick must be thickly buttered with mortar before being butted against the preceding brick (Fig. 3-21).

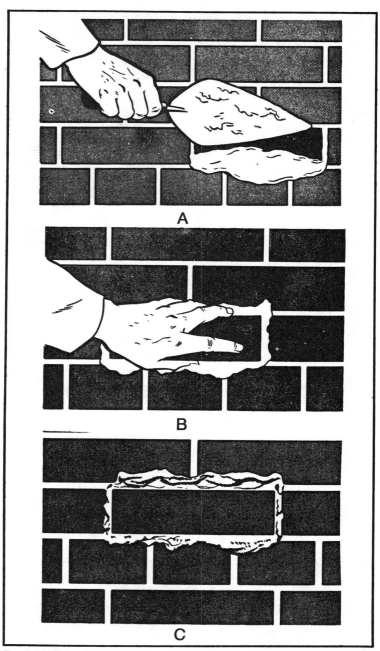

Fig. 3-20. Three steps to inserting a single brick.

Fig. 3-21. Buttering the face of a brick prior to setting it.

Even if the structural integrity of the wall or other unit is not essential, not filling any joint completely with mortar will soon allow excessive penetration of moisture and the resulting loss of the joint. The excessive penetration of moisture will really break up a brick wall in climates that are subject to thaw and freeze cycles—as most areas of this country are. Excess mortar could even squeeze out the top of this header course. Again, any excess on the face and top is cut off (Fig. 3-22).

To finish a header course, you will probably have a single brick gap waiting to be filled. There is not really a problem here, but

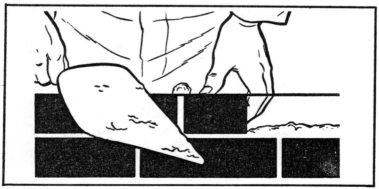
Fig. 3-22. Setting the header course brick in place.

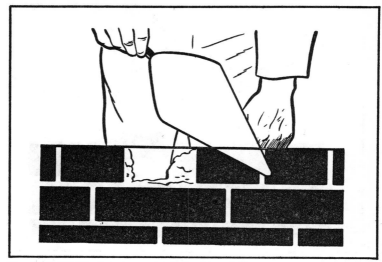

Fig. 3-23. Starting the closure joints for a header course.

instead of buttering the end of the brick you hold in your hand, you butter the ends of the bricks already in place to a thickness of at least one inch (Fig. 3-23). Then place a bed of mortar on the lower course, if it isn't already in place, and insert the closure brick. Try your best not to distrub the bricks already in place (Figs. 3-24 and 3-25).

Fig. 3-24. Getting the brick ready for insertion in the closure.

Fig. 3-25. Setting the brick for the closure.

To close a stretcher course, the closure brick is treated as is the closure brick for a header course, with the exception of one extra step. Butter both ends of the closure brick as thickly as possible, and cover the ends of the bricks already in place. Again, do your best not to disturb the bricks already laid. Any such disturbance means those bricks must be removed, cleaned and relaid. If the bricks are disturbed and not relaid, the bond will be poor and the joints could open a bit and allow water penetration (Figs. 3-26, 3-27, and 3-28).

Fig. 3-26. Preparing to close a stretcher course.

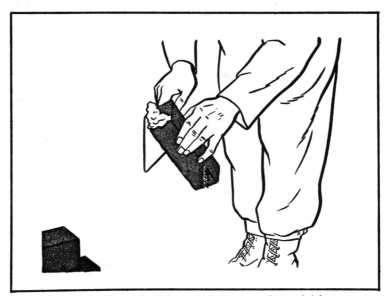
Fig. 3-27. Buttering the ends of the stretcher course closure brick.

One of the most difficult things for the novice bricklayer is discovering how thick the mortar joints should be. Much depends, of course, on the uniformity of the brick being laid since the irregularities of the brick are taken up by the mortar joint. The more regular the brick, the thinner the mortar joint. To a point. Actually,

Fig. 3-28. Setting the closure brick in place.

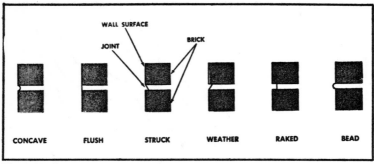

Fig. 3-29. Joint finishes.

optimum bonding, assuming the mortar is correctly mixed, occurs with a quarter-inch joint. With irregular bricks, a joint thickness of up to half an inch is acceptable and will provide good strength.

I've seen some people pull a little trick to fill in header joints that don't seem quite full after the brick is first laid. Mortar is thinned and slushed down into the joints. These slushed joints, as they are called, are bad in several ways. First, the bond is terrible, as the mortar cannot be compacted against the ends of the brick. Second, very often the slushed mortar will run down the faces of already laid brick courses, leaving hard to remove streaks.

Immediately after each course is laid, pick up your jointing tool and compact the mortar in the joints by the process called *pointing*. Pointing consists of gently forcing the tool into the joint and drawing it along to compact the mortar even more than it already has been. In some cases, extra mortar must be added to get a clean looking joint (Fig. 3-29). (When old brick walls are done over, the job is also called pointing. This will be covered in more detail in the chapter on maintaining brick masonry.)

CUTTING BRICK

Often a brick must be cut, as in the case of half-brick for header courses, quarter-brick use in English bonds and so on. A brick chisel is the most accurate tool for any such cutting. The brick is scored with a scribe at the cutting point. The chisel (or bolster) is set on the scribed line and then given a sharp rap with a mason's hammer. Keep the straight side of the cutting edge of the chisel facing the portion of the brick that is to be saved and used (Fig. 3-30).

Brick hammers also have cutting edges for brick since some brick is simply too hard to be cut easily and accurately with a single, neat blow using the brick chisel. The brick hammer is used to chip

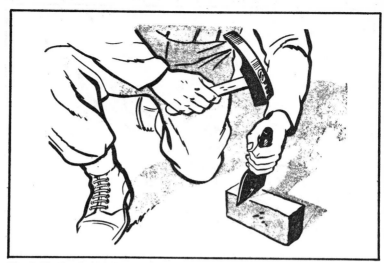

Fig. 3-30. Cutting brick with a brick chisel.

away the brick until the scored line is reached. The brick chisel is then used for the final, neat, cut (Figs. 3-31 and 3-32).

A masonry cut-off wheel, designed to fit either an electric drill or an electric saw, might seem to be the simplest and most rapid

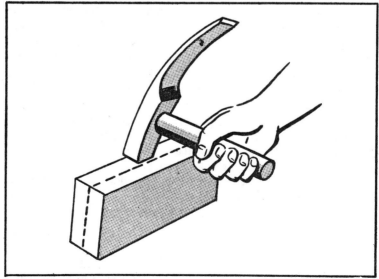

Fig. 3-31. Using a brick hammer to cut brick.

Fig. 3-32. Trimming rough spots on the cut.

method of cutting brick. However, in my experience it is about the worst method anyone could possibly use. For cutting any kind of masonry, I would avoid such a tool whenever possible. First, the quality of the arbors on such blades is almost uniformly poor. Halfway through the first simple cut you're likely to find the blade spinning freely about the chuck of the drill or the arbor shaft of the saw. Second, you must wear a protective mask. If the arbor holds up, the amount of fine dust tossed around could easily prove a definite health hazard. As an experiment—and one I wish we had never tried—my brother and I used such a cut-off blade to shorten a flue pipe. The work, to our great misfortune, took place indoors. The result was a three day cleaning job to get all the dust out of nooks and crannys, plus an overheated half-inch drill. I've never before managed to overheat that drill, even taking a half inch drill bit through thick sheet steel plate. The cut was jagged, as the arbor wore and made clean control nearly impossible. An investment of nearly four dollars for the blade was, in my opinion, a total waste. Whenever possible, and it should almost always be possible in new construction, do all cutting before installation. Use a brick hammer and a brick chisel. We very likely could have done a better job on the flue pipe with one of the little loop saws sprinkled with diamond dust or something similar. And with less overall effort by far.

JOINT FINISHING

There are two reasons to finish joints in exterior masonry work. A tooled joint looks better than a rough one. Second, it compacts the joint mortar and helps to make the joint waterproof. Actually, joints can be cut flush with the faces of the bricks, but the compacting action is lost. As soon as the work dries, you'll begin to see cracks between the bricks and mortar. Generally, the cracks don't extend very far into the joint, but the action of water freezing the thawing can soon extend the distances and reduce your wall to rubble.

Since most people work at a slower pace than all but the slowest professional brick mason, you will need to tool the joints as you build the structure. The mortar joint compresses best and finishes more easily before the joint is set up very much and while the mortar is still rather plastic. Usually a jointing tool is used for the purpose, although other items, depending on the shape you want or need, can be used. At times I've used a finger tip. This is not really recommended for more than about three feet of joint unless you're trying to remove your fingerprints. At other times, the base end of a toothbrush handle has given the shape needed. Experiment a bit if you don't care for the joint shape your jointing tool gives you.

Essentially, there are three types of joints made in brick masonry beyond the unpacked cut-off joint. The concave joint is the best joint from the standpoint of sealing the mortar against water penetration and the effects of the weather. First cut off all excess mortar with your trowel. Now you will need to select the jointing tool made for the purpose. There are several shapes available, from the square ended jointing tool, to a V-ended, to a jointing tool with a curved end. It is best if the tool tip is slightly wider than the joint being worked. Use just enough force to press the mortar tightly against the brick on both sides of the joint and carry your movement along the joint as smoothly as possible. I prefer to do the vertical joints first since that tends to make a cleaner line along the horizontal joints in most kinds of bonds. You might prefer a more broken look.

The flush joint, similar too but not exactly like the cut off joint, means cutting off excess mortar. Use the tip of the trowel at a position slightly away from parallel with the face of the joint. It should be about 10 to 15 degrees, depending on the effect you prefer. Draw it along the joint. This joint is not a stightly packed as the concave joint. Therefore, it is less resistant to water penetration.

For a weather joint, excess mortar is cut away with the trowel. Then the joint is formed by pushing down on the remaining mortar with the top edge of the trowel. While this is effective at shedding running water—that is water running down the face of the brick—it is a much worse joint finish to use in situations where water in large quantities might be blown up against the face of the wall.

The most effective, and to me the most attractive, joint is the concave joint. It is also usually the easiest to make since the jointing tool is specifically designed to provide the style joint you're looking for. All you need do is make sure the mortar is still workable, press the tool in with the appropriate force to get a good mortar pack and draw the tool as steadily as possible. That's it. You're done.

The methods previously described present some of the basics of working with brick. From here on things will get a bit more complex, as you try to work out a good mortar mix and find out what's needed for the base, or footing, under most types of brickwork. You must know not only what happens with mortar, but what sort of concrete, and possibly concrete block is needed to support the final brick structure

4

Concrete & Mortar

You need to know at least a bit about concrete before moving on to mortar and brick laying on a more extensive basis. The reasons are quite simple. First, portland cement makes up a part of mortar. Second, concrete either is used for footings for brickwork or is used for footings for concrete block which will then support brickwork.

Concrete in its usable form has plasticity. It is readily molded, but changes shape rather slowly when the mold is removed. The speed of that shape change is influenced by the mix of the concrete and the quality and character of the finished product is strongly effected also. Also, the degree of plasticity affects the workability of the concrete. Concrete with a very stiff mix, as shown by a slump test (explained in a short while), is needed for certain types of work. It would be totally out of place where the concrete mix had to flow into heavily reinforced sections or into smaller openings in forms. Workability is primarily controlled by the amounts and proportion of fine to coarse aggregates used with a given quantity of concrete paste (water and pure concrete).

Uniformity of mix is essential for greatest overall strength. Each batch should meet the same specifications does the first.

Naturally, strength is of great importance. One of the basic reasons for using concrete is its great compressive strength. The ratio of water to cement used is the basic factor influenciing final mix strength. However, other variables must also be controlled. Hydration (drying) time is partly controlled by the amount of water added

Table 4-1. Water/Cement Ratios.

| Type of structure | Severe wide range in temperature or frequent alternations of freezing and thawing (air-entrained concrete only) (gallons/sack) | | | Mild temperature rarely below freezing, or rainy, or arid (gallons/sack) | | |
| | In air | At water line or within range of fluctuating water level or spray | | In air | At water line or within range of fluctuating water level or spray | |
		In fresh water	In sea water or in contact with sulfates†		In fresh water	In sea water or in contact with sulfates†
Thin sections such as reinforced piles and pipe	5.5	5	4.5	6	5.5	4.5
Bridge decks	5	5	4.5	5.5	5.5	5
Thin sections such as railings, curbs, sills, ledges, ornamental or architectural concrete, and all sections with less than 1-in. concrete cover over reinforcement	5.5	....	....	6	5.5	....
Moderate sections, such as retaining walls, abutments, piers, girders, beams	6	5.5	5	††	6	5
Exterior portions of heavy (mass) sections	6.5	5.5	5	††	6	5

Exposure condition				
Concrete deposited by tremie under water	5	5	5	5
Concrete slabs laid on the ground	6	....	....	5
Pavements	5.5	....	....	....
Concrete protected from the weather, interiors of buildings, concrete below ground	††	††	6	....
Concrete which will later be protected by enclosure or backfill but which may be exposed to freezing and thawing for several years before such protection is offered	6	††	....	....

*Adapted from Recommended Practice for Selecting Proportions for Concrete (ACI 613-54).

**Air-entrained concrete should be used under all conditions involving severe exposure and may be used under mild exposure conditions to improve workability of the mixture.

†Soil or groundwater containing sulfate concentrations of more than 0.2 percent. For moderate sulfate resistance, the tricalcium aluminate content of the cement should be limited to 8 per cent, and for high sulfate resistance to 5 percent. At equal cement contents, air-entrained concrete is significantly more resistant to sulfate attack than non-air-entrained concrete.

††Water-cement ratio should be selected on basis of strength and workability requirements, but minimum cement content should not be less than 470 lbs. per cubuc yard.

at the outset. It can also be partly controlled by keeping the mix damp and "curing" for varying lengths of time. Each sack of cement needs about 2 ½ gallons of water to obtain the proper chemical drying action that takes place when cement cures. The thinner the paste after this point, the weaker the final result. Generally, each sack of cement requires a minimum of 4 gallons of water and a maximum of 8 gallons. The amount depends on just how damp your fine and coarse aggregates—sand and gravel—are when mixing takes place. Simply dumping in so many gallons of water per specified ratio of dry mix doesn't always work exactly.

For greatest water tightness, the Army Corps of Engineers recommends using no more than 6 gallons of water per sack of cement. Durability decreases when too much water is used in the mix (Table 4-1).

Portland cements are mixtures of raw materials, finely ground and heated to a fusion temperature of about 2700 degrees F and again finely ground. When combined with water, Portland cements harden through hydration to form a rock like mass. In general, most home concrete work will be done with ASTM Type I Portland Cement. The American Society for Testing Materials provides categories for other types of Portland cements, but they are of little general interest. ASTM Type I Portland cement is classified as a general cement for use where special properties such as high resistance to certain acids or low heat drying are not needed. Uses include pavement of all kinds, reinforced concrete buildings, bridges, tanks, culverts, masonry units, soil cement mixtures, mortars and other such things.

About the only requirement for storing portland cement is to keep it dry. If it is kept dry, the cement will last indefinitely. Moisture contact can either set it into a mass, or cause it to set more slowly when used, with lower resulting strength.

Water used in making cement should be as pure and clean as possible. If your area is one of those where excessive amounts of sulfates are found, the water should not be used even though it is classified as fit to drink. However, in most cases if the water is classified as fit to drink it can be used to make cement.

AGGREGATES

The characteristics of aggregates—which make up from sixty to eighy percent of the final dry mix—used in concrete work have a large affect on the final product. Rough textured aggregates require more water to produce workable concrete than do smoother particles. With rough aggregates, more cement must also be added to keep the water/cement ratio in balance for the correct final strength.

Aggregates range from fine mason's sand to large gravel, crushed stone, cinders, and occasionally, burned clay. The result, with a proper mix, is concrete that weighs from 135 to 160 pounds per cubic foot. Aggregates should be as free of contamination as is possible. Some types can be washed to rid them of dirt, organic material, salts and other possible contaminates, sand which contains organic materials cannot be decontaminated. In most areas of the country, simply buying a good mason's sand will provide you with the qualities needed. The same is true for good bank run gravel that is washed and sized for a particular job. At times and in some areas, good aggregate can be a bit hard to come by and more expensive than seems reasonable. It simply doesn't make sense to do your job with less than the best materials available for that job if you wish to have excellent qualities in the finished work (Table 4-2).

Aggregates are graded for abrasion resistance, resistance to freezing and thawing and chemical stability. Particle shape and texture also assume some importance, as does particle size. Generally, you're safe using whatever type of sand or gravel the local contractors in your area prefer. Most reputable contractors will be looking for good quality as well as lowest cost consistent with that quality (Fig. 4-1).

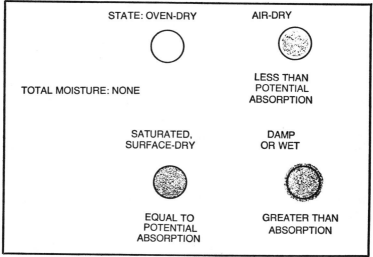

Fig. 4-1. Moisture conditions of aggregates.

Grading for size depends on the type work you are doing. When, as with brick masonry, your major concern is a footing or a footing and base wall to support brick work, the largest coarse aggregate used shouldn't be more than a quarter the thickness of the unit being constructed. In other words, if you are pouring an 8-inch footing, the largest piece of coarse aggregate should be no more than 2 inches across. In no case should course aggregate exceed 2 ½ inches in size.

With aggregates of any kind, asssuming the basic quality is there, cleanliness is most important. If you have any doubts about the cleanliness of your aggregate, it is a simple matter to run a quick test for silt. Take a quart jar and place about 2 inches of the aggregate in the bottom. Fill the jar with water and give the jar a gentle shake. Stand it aside until the aggregate settles and then check the depth of the silt on top of the aggregate. If you find more than about one-eighth of an inch of silt, the aggregate will need to be washed before use. I use a simple process for this washing, but those without a pick-up truck will have to find some other container. I park the loaded pick-up on a grade, tailgate facing the lower slant of the grade, but still up. Then any hose from a source of clean water can be used to soak the aggregate until the water running from around the tailgate is clear, or reasonably clear. Since my half-ton Dodge, even with booster springs and special shocks, is overloaded long before its 8-foot bed is half-full of most aggregates, I'm working with a layer no more than a foot or so thick. The washing doesn't take too long. At this point, I usually shovel the aggregate out on as wide a surface as possible and let it dry for a couple of days—praying for no rain, of course, but running out to cover the stuff when the sky clouds up. Letting the aggregate dry is not always essential for some types of concrete, but it is almost always essential for mortar use since sand that is too wet will form a slushy mixture.

Testing for organic material in sand and other aggregates is sometimes needed. Again, you'll need a glass jar into which you place about one measuring cup of the suspect aggregate. If the aggregate is sand and you suspect the presence of organic material, do the test before making the purchase. Add a cup of clean water and one heaping teaspoon of household lye. Don't use the type of lye in drain cleaners such as Drano as these have aluminum chips added that interfere with the testing. Use just plain, ol' lye and nothing else. Shake well, though gently, until the lye is completely dissolved.

Table 4-2. Gradation Needs for Coarse Aggregates.

Size Number	Nominal size (sieves) with square openings	Amount finer than each laboratory sieve (square openings), percent by weight						
		4 in.	3½ in.	3 in.	2½ in.	2 in.	1½ in.	1 in.
1	3½ to 1½ in	100	90 to 100		25 to 60		0 to 15	
2	2½ to 1½ in			100	90 to 100	35 to 70	0 to 15	
357	2 in. to No. 4				100	95 to 100		35 to 70
467	1½ in. to No. 4					100	95 to 100	
57	1 in. to No. 4						100	95 to 100
67	¾ in. to No. 4							100
7	½ in. to No. 4							
8	⅜ in. to No. 8							
3	2 to 1 in				100	90 to 100	35 to 70	0 to 15
4	1½ to ¾ in					100	90 to 100	20 to 55

Size Number	Nominal size (sieves) with square openings	Amount finer than each laboratory sieve (square openings), percent by weight					
		¾ in.	½ in.	⅜ in.	No. 4 (4760—micron)	No. 8 (2380—micron)	No. 16 (1190—micron)
1	3½ to 1½ in	0 to 5					
2	2½ to 1½ in	0 to 5					
357	2 in. to No. 4		10 to 30		0 to 5		
467	1½ in. to No. 4	35 to 70		10 to 30	0 to 5		
57	1 in. to No. 4		25 to 60		0 to 10	0 to 5	
67	¾ in. to No. 4	90 to 100		20 to 55	0 to 10	0 to 5	
7	½ in. to No. 4	100	90 to 100	40 to 70	0 to 15	0 to 5	
8	⅜ in. to No. 8		100	85 to 100	10 to 30	0 to 10	0 to 5
3	2 to 1 in	0 to 5					
4	1½ to ¾ in	0 to 15	0 to 5				

*From specifications for concrete aggregate (ASTM—C33).

Set the jar aside for about 24 hours and then take a look at the color of the contents. If the color is orange, there is organic material present. If it is more than a very, very pale orange, the aggregate must be washed. As I previously explained, sand that is contaminated with organic material cannot be washed. This is why suspect sand should be tested before purchase.

Try to purchase all aggregates by cubic measures, either cubic feet or yards (usually cubic yards). In some areas—I now live in one—aggregates are still sold by weight. This is a good way to get ripped off. If you're not doing business with a reputable company or person, anyone selling such material by weight can add considerably to the profit margin by the simply hooking up a garden hose and soaking down the load. I've seen many dump trucks loaded with gravel running along old country roads with water literally streaming from the corners of the bed. Even with honest aggregate merchants, the weight of the aggregate can vary greatly from day to day, depending on its moisture content.

Even when aggregates are sold by volume, the merchant can try bulking by adding so much water that the individual particles (this works best with very fine aggregates) are held apart to increase the load volume. Fortunately, the amount of water needed about equals the extra cost to you, so not too many dealers will try this one unless they've got a river handy. Anytime you notice water dripping from aggregate, whether you're buying by volume or by weight, you should cast a wary eye on the dealer. In fact, you should most likely take your business elsewhere unless there is a good explanation and some allowance made for the water weight or volume.

CONCRETE MIXTURES

The three methods of determining the mixture proportions for concrete are book, trial and absolute volume. In almost every case, the first two methods must be combined since the book method is really theoretical and must be adjusted, after testing, in the field. The absolute volume method also requires a slump test—something few do-it-yourselfers seem to bother with—but a relatively simple way for the novice to tell whether or not the mixture is within needed limits. Volume methods are most often used in home applications.

First you'll need to know the qualities of the concrete needed for a particular job and then you can mix to suit those qualities. For our purposes, the usual maximum allowable slump will be 6 inches,

Table 4-3. Recommended Slumps.

Type of construction	Slump, inches	
	Maximum	Minimum
Reinforced foundation walls and footings	6	3
Unreinforced footings, caissons, and sub-structure walls	4	1
Reinforced slabs, beams, and walls	6	3
Building columns	6	4
Pavements	3	1
Heavy mass construction	3	1
Bridge decks	4	3
Sidewalk, driveway, and slabs on ground	6	3

*When high-frequency vibrators are used, the values may be decreased approximately one-third, but in no case should the slump exceed 6 inches.

with a minimum of 3 inches, if the footings are reinforced (steel bars). For non-reinforced footings, the maximum allowable slump is 4 inches, with a minimum of 1 inch (Table 4-3).

SLUMP TEST

To do a slump test you will need a slump cone, which you will probably have to make yourself since they can be a bit hard to locate. If that's necessary, the cone is best made of galvanized metal, in 16-gauge weight. It is 12 inches high, with a base 8 inches in diameter and an open top 4 inches in diameter. You'll also need a two-foot long, smooth, pointed steel bar as a tamper. The cone can be made relatively easily with tin snips, an electric drill and a pop rivet gun (Fig. 4-2).

Take samples for the slump test from the batch you're mixing as soon as the final wet mix is completed. After wetting the cone well, place a flat board under the testing cone. Fill the cone in three layers. Make each as close as possible to one-third the cone's volume as you can. Move the scoop or shovel around the edge of the cone instead of just dumping the concrete in, so that you get a symmetrical layer each time. Using the tamping bar, rod in each layer immediately after placing it in the cone. Try to distribute your strokes over the surface of the mix as evenly as possible. Use 25 strokes to each layer. Penetrate the just laid layer into the one below (you should hit the board on the first layer). Once the cone is filled a bit higher than its top, strike off the excess with a straightedge and immediately lift the cone, straight up and gently, from the mix. You then measure the slump, using the cone as a guide, as Fig. 4-3 shows.

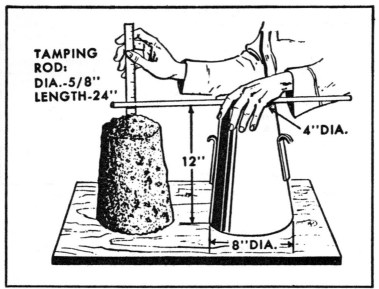

Fig. 4-2. Measurement of slumps.

A final check of the slump tested mix will help to show its workability and cohesiveness. Gently tap the side of the slumped cone with the tamping rod. If the mix is going to be easily workable, with good cohesiveness, it will merely slump a little bit lower. If, instead, it crumbles or the aggregates separate, the mix is incorrect. Crumbling indicates too much sand in the mixtures, while separation of the aggregates indicates too little sand.

WORKABILITY

Workability is an important factor with concrete, for a properly workable mix fills all form spaces completely. A slushy mixture poured into a form will do the same, but the over use of water in such mixes weakens the concrete after curing takes place. Ideally, when working out a water to cement ratio only enough water should be used to start hydration. A compromise is needed, since the resulting mix would be far too stiff for almost all applications. The slump test determines the workability. The more slump, the greater the workability. If you have fairly complex forms, where a lot of flow is needed into small area, then go to the lower end of the slump so the concrete will flow easily.

Using too much sand or too little sand is cured by adding more cement, in the first case, and more sand in the second case. Extra

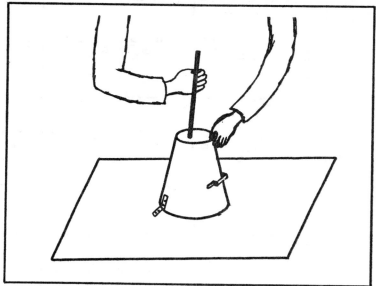

Fig. 4-3. Charging a slump test cone.

water can also be used, but if you add extra water to the concrete to increase workability, then you must also add extra cement to maintain the cement to water ratio.

As a general rule, you can figure that adding extra fine aggregate to a batch to change things will result in a stiffer mix, while adding extra coarse aggregate will loosen it a bit.

Tables 4-4, 4-5, 4-6 and 4-7 will give you a starting point for working out cement to water to aggregate mixtures for a variety of strengths. Those at the bottom of the Tables are the weakest and those at the top of the least workable.

MIXING

Cement comes in 94 pound bags with a volume of just about one cubic foot. All other portions of the mix are worked from this point, with a more or less standard for home use proportion of one part of cement to two parts of sand and three of aggregate under 1 ½ inches in size. For this, all measuring is done by volume, not by weight. For footings and walls, because they don't require resistance to abrasion and water erosion, the proportions would be one part cement, two parts sand and four of aggregates under 1 ½ inches in size.

While those are standards, you should consider a fact I have already mentioned about sand: bulking. Adding an extra quarter

Table 4-4. Trial Mixes.

Water-cement ratio Gal per sack	Maximum size of aggregate inches	Air content (entrapped air) per cent	Water gal per cu yd of concrete	Cement sacks per cu yd of concrete	With fine sand—fineness modulus = 2.50		
					Fine aggregate per cent of total aggregate	Fine aggregate lb per cu yd of concrete	Coarse aggregate lb per cu yd of concrete
4.5	3/8	3	46	10.3	50	1240	1260
	1/2	2.5	44	9.8	42	1100	1520
	3/4	2	41	9.1	35	960	1800
	1	1.5	39	8.7	32	910	1940
	1½	1	36	8.0	29	880	2110
5.0	3/8	3	46	9.2	51	1330	1260
	1/2	2.5	44	8.8	44	1130	1520
	3/4	2	41	8.2	37	1040	1800
	1	1.5	39	7.8	34	990	1940
	1½	1	36	7.2	31	960	2110
5.5	3/8	3	46	8.4	52	1390	1260
	1/2	2.5	44	8.0	45	1240	1520
	3/4	2	41	7.5	38	1090	1800
	1	1.5	39	7.1	35	1040	1940
	1½	1	36	6.5	32	1000	2110
6.0	3/8	3	46	7.7	53	1440	1260
	1/2	2.5	44	7.3	46	1290	1520
	3/4	2	41	6.8	39	1130	1800
	1	1.5	39	6.5	36	1080	1940
	1½	1	36	6.0	33	1040	2110

6.5	3/8	3	46	7.1	54	1480	1260
	1/2	2.5	44	6.8	46	1320	1520
	3/4	2	41	6.3	39	1190	1800
	1	1.5	39	6.0	37	1120	1940
	1½	1	36	5.5	34	1070	2110
7.0	3/8	3	46	6.6	55	1520	1260
	1/2	2.5	44	6.3	47	1360	1520
	3/4	2	41	5.9	40	1200	1800
	1	1.5	39	5.6	37	1150	1940
	1½	1	36	5.1	34	1100	2110
7.5	3/8	3	46	6.1	55	1560	1260
	1/2	2.5	44	5.9	48	1400	1520
	3/4	2	41	5.5	41	1240	1800
	1	1.5	39	5.2	38	1190	1940
	1½	1	36	4.8	35	1130	2110
8.0	3/8	3	46	5.7	56	1600	1260
	1/2	2.5	44	5.5	48	1440	1520
	3/4	2	41	5.1	42	1230	1800
	1	1.5	39	4.9	39	1220	1940
	1½	1	36	4.5	35	1160	2110

Table 4-5. Trial Mixes.

Water-cement ratio Gal per sack	With average sand—fineness modulus = 2.75			With coarse sand—fineness modulus = 2.90		
	Fine aggregate percent of total aggregate	Fine aggregate lb per cu yd of concrete	Coarse aggregate lb per cu yd of concrete	Fine aggregate percent of total aggregate	Fine aggregate lb per cu yd of concrete	Coarse aggregate lb per cu yd of concrete
4.5	52	1310	1190	54	1350	1150
	45	1170	1450	47	1220	1400
	37	1030	1730	39	1080	1680
	34	980	1870	36	1020	1830
	32	960	2030	33	1000	1990
5.0	54	1400	1190	56	1440	1150
	46	1250	1450	48	1300	1400
	39	1110	1730	41	1160	1680
	36	1060	1870	38	1100	1830
	34	1040	2030	35	1080	1990
5.5	55	1460	1190	57	1500	1150
	47	1310	1450	49	1360	1400
	40	1160	1730	42	1210	1680
	37	1110	1870	39	1150	1830
	35	1080	2030	36	1120	1990
6.0	56	1510	1190	57	1550	1150
	48	1360	1450	50	1410	1400
	41	1200	1730	43	1250	1600
	38	1150	1870	39	1190	1830
	36	1120	2030	37	1160	1990

6.5	57	1550	1190	58	1590	1150
	49	1390	1450	51	1440	1400
	49	1390	1450	51	1440	1400
	42	1240	1730	43	1290	1680
	39	1190	1870	40	1230	1830
	36	1150	2030	37	1190	1990
7.0	57	1590	1190	59	1630	1150
	50	1430	1450	51	1480	1400
	42	1270	1730	44	1320	1680
	39	1220	1870	41	1260	1830
	37	1180	2030	38	1220	1990
	58	1630	1190	59	1670	1150
	50	1470	1450	52	1520	1400
	43	1310	1730	45	1370	1600
	40	1260	1870	42	1300	1830
	37	1210	2030	39	1250	1990
8.0	58	1670	1190	60	1710	1150
	51	1520	1450	53	1560	1400
	44	1350	1730	45	1400	1680
	41	1290	1870	42	1330	1830
	38	1250	2030	39	1280	1990

*Increase or decrease water per cubic yard by 3 per cent for each increase or decrease of 1 in. in slump, then calculate quantities by absolute volume method. For manufactured fine aggregate, increase percentage of fine aggregate by 3 and water by 17 lb. per cubic yard of concrete. For less workable concrete, as in pavements decrease percentage of fine aggregate by 3 and water by 8 lb per cubic yard of concrete.

Table 4-6. Trial Mixes.

Water-cement ratio Gal per sack	With average sand—fineness modulus = 2.75			With coarse sand—fineness modulus = 2.90		
	Fin aggregate percent of total aggregate	Fine aggregate lb per cu yd of concrete	Coarse aggregate lb per cu yd of concrete	Fine aggregate percent of total aggregate	Fine aggregate lb per cu yd of concrete	Coarse aggregate lb per cu yd of concrete
4.5	53	1320	1190	54	1360	1150
	44	1130	1450	46	1180	1400
	38	1040	1730	39	1090	1680
	34	970	1870	36	1010	1830
	32	950	2030	33	990	1990
5.0	54	1400	1190	56	1440	1150
	46	1210	1450	47	1260	1400
	39	1110	1730	41	1160	1630
	36	1040	1870	37	1080	1830
	34	1050	2030	35	1090	1990
5.5	55	1460	1190	57	1500	1150
	46	1260	1450	48	1310	1400
	40	1150	1730	42	1210	1680
	37	1080	1870	38	1120	1830
	34	1040	2030	35	1090	1990
6.0	56	1500	1190	57	1540	1150
	47	1300	1450	49	1350	1400
	41	1190	1730	42	1240	1680
	37	1110	1870	39	1150	1830
	35	1090	2030	36	1130	1990

Slump	%			%		
6.5	56	1530	1190	58	1570	1150
	48	1330	1450	50	1380	1400
	41	1220	1730	43	1270	1680
	38	1150	1870	39	1190	1830
	36	1120	2030	37	1160	1990
7.0	57	1570	1190	58	1610	1150
	49	1370	1450	50	1420	1400
	42	1250	1730	44	1300	1680
	38	1170	1870	40	1210	1830
	36	1140	2030	37	1180	1990
7.5	57	1600	1190	59	1640	1150
	49	1400	1450	51	1450	1400
	43	1280	1730	44	1330	1680
	39	1210	1870	41	1250	1830
	37	1170	2030	38	1210	1990
8.0	58	1630	1190	59	1670	1150
	50	1430	1450	51	1480	1400
	43	1310	1730	44	1360	1680
	40	1230	1870	41	1270	1830
	37	1190	2030	38	1230	1990

*Increase or decrease water per cubic yard by 3 per cent for each increase of 1 in. in slump, then calculate by absolute volume method. For manufactured fine aggregate, increase percentage of fine aggregate by 3 and water by 17 lb. per cubic yard of concrete. For less workable concrete, as in pavements decrease percentage of fine aggregate by 3 and water by 8 lb. per cubic yard of concrete.

Table 4-7. Trial Mixes.

Water-cement ratio Gal per sack	Maximum size of aggregate inches	Air Content (entrapped air) per cent	Water gal per cy yd of concrete	Cement sacks per cu yd of concrete	With fine sand—fineness modulus = 2.50		
					Fine aggregate per cent of total aggregate	Fine aggregate lb per cu yd of concrete	Coarse aggregate lb per cu yd of concrete
4.5	3/8	7.5	41	9.1	50	1250	1260
	1/2	7.5	39	8.7	41	1060	1520
	3/4	6	36	8.0	35	970	1800
	1	6	34	7.8	32	900	1940
	1 1/2	5	32	7.1	29	870	2110
5.0	3/8	7.5	41	8.2	51	1330	1260
	1/2	7.5	39	7.8	43	1140	1520
	3/4	6	36	7.2	37	1040	1800
	1	6	34	6.8	33	970	1940
	1 1/2	5	32	6.4	31	930	2110
5.5	3/8	7.5	41	7.5	52	1390	1260
	1/2	7.5	39	7.1	44	1190	1520
	3/4	6	36	6.5	38	1090	1800
	1	6	34	6.2	34	1010	1940
	1 1/2	5	32	5.8	32	970	2110
6.0	3/8	7.5	41	6.8	53	1430	1260
	1/2	7.5	39	6.5	45	1230	1520
	3/4	6	36	6.0	38	1120	1800
	1	6	34	5.7	35	1040	1940
	1 1/2	5	32	5.3	32	1010	2110

6.5	⅜	7.5	41	6.3	54	1460	1260
	½	7.5	39	6.0	45	1260	1520
	¾	6	36	5.5	39	1150	1800
	1	6	34	5.2	36	1080	1940
	1½	5	32	4.9	33	1040	2110
7.0	⅜	7.5	41	5.9	54	1500	1260
	½	7.5	39	5.6	46	1300	1520
	¾	6	36	5.1	40	1180	1800
	1	6	34	4.9	36	1100	1940
	1½	5	32	4.6	33	1060	2110
7.5	⅜	7.5	41	5.5	55	1530	1260
	½	7.5	39	5.2	47	1330	1520
	¾	6	36	4.8	40	1210	1800
	1	6	34	4.5	37	1140	1940
	1½	5	32	4.3	34	1090	2110
8.0	⅜	7.5	41	5.1	55	1560	1260
	½	7.5	39	4.9	47	1360	1520
	¾	6	36	4.5	41	1240	1800
	1	6	34	4.3	37	1160	1940
	1½	5	32	4.0	34	1110	2110

portion of sand will counteract any bulking action, and there will be some since sand is almost never totally dry.

Water is an important ingredient of the concrete mix, and can cause a lot of difficulties. There are many formulas and standards for the amount of water required, as you can see from some of the earlier Tables. But for most uses, such extra great care to get an exact water to cement ratio is beyond the needs, as well as the capabilities, or the homeowner. In fact, in most professional uses, the amount of water content of the various aggregates ends up being estimated anyway. Using the slump test, with the tap test at the end, will provide most of the information you need as to the correctness of the mixture. Even using as much as 7 gallons of water to one bag of cement will result in a concrete strong enough for almost all around the home (or under the home) purposes.

As an example, using seven gallons of water to one bag of cement will provide you with concrete having a compressive strength of about 3000 pounds per square inch after a 28 day cure. And no matter what you do to weaken concrete, it will have a low tensile strength (resistance to twisting forces). Low tensile strength—about the highest is not much over 600 pounds per square inch—of concrete is the reason it cracks.

The dry mix should be complete. That is, the sand and coarse aggregates and cement should be thoroughly and completely mixed with one another before water is added.

ESTIMATING AMOUNTS

With all the information on amounts going into a mix, you might be wondering just a bit about how you can tell just how much of the total you'll need to fill your footing forms. Essentially, all you need do is find the volume, in cubic yards, of the form to be filled and work from there. Unfortunately, most such figuring must be done in inches. As an example, it is reasonably normal to make a footing 16-inches wide. On normal undisturbed soil, footings are generally built twice the width they support and the average concrete block is 8 inches wide. Assume a 12-inch depth for the footing, and a 20-foot length. Your first job is to determine the volume of the footing in cubic inches: $16 \times 12 \times 240 = 46080$ cubic inches. To reduce this to cubic feet, you divide by 1,728 to get 26.66666. Since a cubic yard contains 27 cubic feet, this footing would require just about 1 cubic yard.

There is a certain amount of sloppiness when working with concrete. A percentage should be added to allow for material lost because the bottom of the footing isn't perfectly flat, the amount that sticks to the container and so on. If your original figures come out under four or five cubic yards, it's wise to add 15 percent to your total. Anything over 5 cubic yards needs a 10 percent waste allowance.

Ready mix concrete generally comes in 94 pound or one cubic foot, sacks. For brick masonry this is usually impractical, since you'll be using concrete for footings and other moderately large sections. It tends to be a lot more expensive than concrete you mix yourself or have delivered mixed and ready to pour. Still, if you need to lay down a simple footing for a small project, you might want to use ready mix. The most important consideration with such a mix is the use of water. Add too much and the odds will be excellent that you won't have any sand, cement or other material to get the proper degree of wetness. In most cases, you'll need less than a gallon and the water should be added in small amounts, starting with no more than two quarts.

FORMS

Occasionally, a footing can be poured directly in a neatly cut trench or hole in the ground, but in most cases some type of form must be used. Depending on complexity, the cost of forms for concrete work could be as much as a third of the total cost of the project (for the concrete work, not the brick work). You'll find it pays to design and build carefully. Sloppy or weak forms can result in miserably inadequate bases for the later brickwork. I've seen over optimistic people think that expanses 16 feet long and 4 feet high, to a width of about 2 feet, needed only sheets of three-quarter inch plywood with cats extending to the sides of the trench every couple of feet along the length. The result is always the same; at best a badly bowed foundation wall. Often the form will completely collapse in several spots, forcing the addition of more concrete mix to almost fill in the trench. This adds wildly to the overall cost since a collapsed form can ruin, depending on the job size, several cubic yards or even more.

Most forms are for slabs or footings or the occasional foundation wall to support the brick work. When it isn't possible to use the original trench, the form will usually be of wood. Metal forms that are used many times are too expensive for one time use around the home. In any case, even the professionals use wooden forms since the lumber, if properly handled, can often be used for other building purposes later on in the project.

Oddly enough, lumber used for concrete forms not only need not be the more expensive kiln dried types, but should not be more than partially seasoned. Kiln dried wood will absorb much of the water from the concrete mix and swell, allowing bulging and distortion. If you use totally green lumber, keep it wet until the concrete is poured so that shrinkage will be minimal and cracks won't open to allow concrete to escape. For footing and foundation work, rough cut limber is fine since the walls will not be visible later on and a smooth surface isn't needed. Lumber that is used for the portions of concrete work that will later be visible should have at least the side that faces the inside of the form planed.

If plywood is used for forms, as is often the case when large expanses are to be built, it must be made with waterproof glue. For this reason specify one of the exterior types. You do not need anything better than a CD classification in either a five-eighth inch or three-quarter inch thickness. The larger the form, the thicker the plywood must be.

Form design involves building in enough strength to hold the concrete as it is poured in a plastic state and until it has hardened. There are three variables affecting the strength needed in your forms. The mixer output is important because a mixer that can dump a lot of concrete at one time builds up a greater head of pressure and requires a stronger form. The area to be enclosed within the forms determines the total weight of the concrete to be dumped. The ambient temperature determines the time the concrete will take to set up and affects the time duration durability of the forms. At about 70 degrees F, concrete will take its inital set in about 90 minutes or so. Higher temperatures have a slightly faster set and lower temperatures have a slower set.

If you're using a hand mixer or a motor driven small mixer, you will need to know the mixer yield in cubic feet and the time needed to prepare each batch. The following formula will provide the mixer output figure for the final equation:

$$\frac{\text{mixer yield (cubic feet)}}{\text{batch time (minutes)}} \times \frac{\text{60 minutes}}{\text{hour}}$$

$$= \text{mixer output (cubic feet per hour)}$$

The area of the form is found by multiplying the length times the width. The temperature is estimated or read off a thermometer.

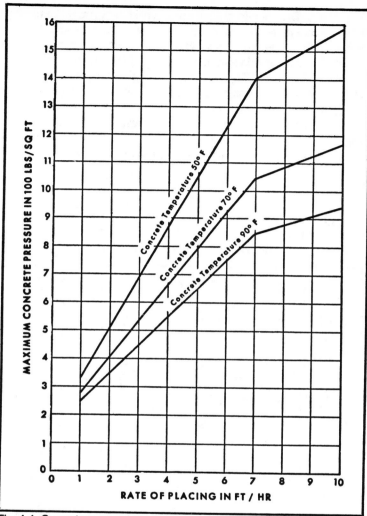

Fig. 4-4. Concrete pressure graph.

To get the rate of placement per hour, divide the mixer output in cubic feet per hour by the plan area in square feet. Then, using Fig. 4-4, enter the rate of placement at the proper place on the chart and draw a vertical line to intersect the nearest to correct temperature line. Read horizontally across to the left then to find your maximum concrete pressure.

Use the maximum concrete pressure figure to determine maximum stud spacing by entering it on the bottom of the chart in

Figure 4-5. Draw a vertical line up the chart until you reach the type of sheathing material you are going to use. Again, read to the left, horizontally, to get your maximum stud spacing for supports. If the answer runs on an odd number, for example 25.5 inches, then drop it back to 24 inches. This not only provides a margin of safety, but makes the building a bit simpler as you will be working with more easily measured units.

The next job is to determine the uniform load on each stud. The formula is:

maximum concrete pressure (pounds/square feet)

×

stud spacing (feet)

=

uniform load on stud (pound per lineal foot)

This figure is then entered on the bottom line of the chart in Fig. 4-11 and a line is drawn vertically until it cuts across the correct

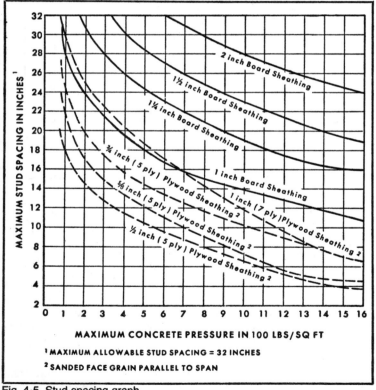

Fig. 4-5. Stud spacing graph.

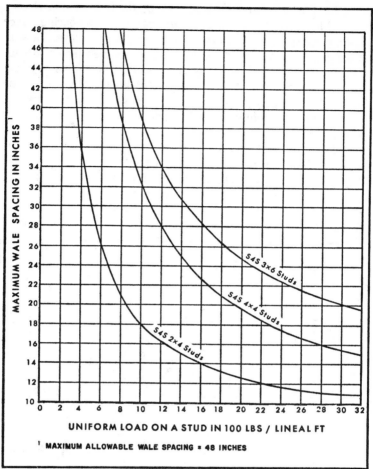

Fig. 4-6. Stud load graph.

stud size curve. Again, a horizontal reading will give the correct maximum wall spacing, which is once more rounded down to an even number of inches if it doesn't fall on such a figure.

For larger wall areas, tie wires might be needed. Their maximum spacing must also be determined. The uniform load on a wall is needed to figure tie wire spacing. The formula is:

maximum concrete pressure (pound/square feet)

×

wall spacing (feet)

=

uniform load on a wale) (pounds per lineal foot)

77

See Fig. 4-7 and enter the figure on the bottom line. Draw a vertical line and read horizontally to get maximum spacing. Once more, if the spacing is not an even number, round it down. Tie wire spacing is also based on tie wire strength. See Table 4-8 for information from the U.S. Army supply system for minimum breaking load for types of tie wire where the strength is unknown. The tie wire spacing in inches equals tie wire strength in pounds times 12 (inches per foot) with that result divided by the uniform load on a wall (pounds per foot).

If the tie wire spacing is shorter than the maximum stud spacing, reduce the maximum stud spacing to the tire wire spacing. If the maximum tie wire spacing is greater than the stud spacing, tie at the intersections of the studs and walls.

With all the design details at hand, the actual construction of forms becomes relatively simple. Make any excavation needed—

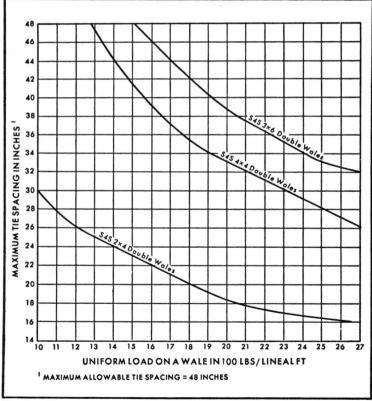

Fig. 4-7. Wale load graph.

Table 4-8. Breaking Load of Wire.

STEEL WIRE	
Size of wire Gage No.	Minimum breaking load double strand Pounds
8	1700
9	1420
10	1170
11	930
BARBED WIRE	
Size of each wire Gage No.	Minimum breaking load Pounds
12½	950
13[1]	660
13½	950
14	650
15½	850
[1]Single strand barbed wire.	

always below the local frost depth for footings and foundation walls. Nail sizes will vary depending on the material sizes, but make certain you have good penetration for maximum holding power. For basic brickwork directly on a footing, keying the footing (laying in a beveled 2 × 4 to form a key) is not needed. But if a concrete wall will be built on the footing and later faced in brick, keying is a good idea for increased strength.

Oil the 2 × 4 before inserting the concrete at the top of the footing. Form panels will have less concrete sticking to them if they are also oiled before pouring, though a coat of pain will do almost as good a job and might make the lumber more suitable for later use in roofing, sheathing or some other place.

Excavation for footings should be no less than the distance specified by your local building codes. This is often the frost level. If the earth is disturbed or soft and easily moved, greater depth to reach firm earth might be needed. One other solution for footings on wet or soft earth, to help prevent shifting and cracking of the concrete and any structure built on the footing, is to increase the size of the footing so that it "floats" its load over a wider area. Moisten the earth at the bottom of the forms before pouring the concrete (Figs. 4-8, 4-9, 4-10, 4-11, 4-12, 4-13 and 4-14).

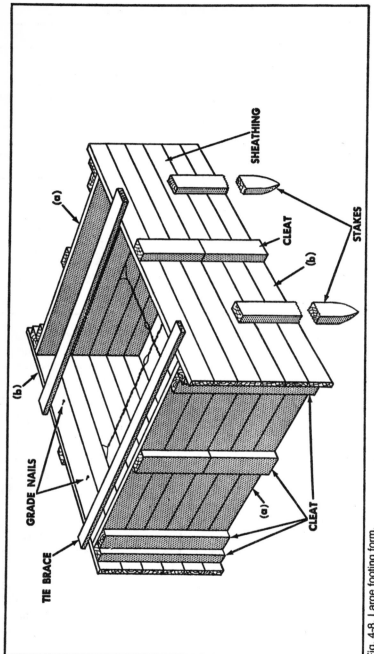

Fig. 4-8. Large footing form.

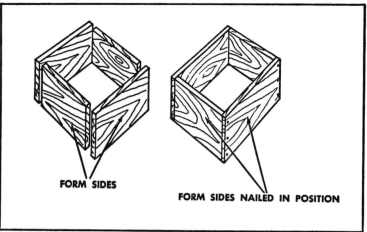

Fig. 4-9. Small footing form.

The concrete is deposited in horizontal layers as evenly as you can possibly do the job. The layers should be no less than 6 inches in depth, with each succeeding layer poured before the layer ahead of it has time to take its initital set.

A screed—a straight edged board—is used to strike off excess concrete at the tops of the forms.

Once the concrete is in place, curing will begin. As I said, concrete cures by a chemical process known as hydration. To get a good cure, concrete must be kept moist. There are two basic ways to do this. One is to prevent the loss of water from the concrete after it is poured. Water is essential to hydration and once the water content drops below certain levels, hydration, and therefore curing, stops. The second method is to add water to the surface after pouring.

The best all around curing method is to lay plastic sheeting over the concrete. As Table 4-9 indicates, there are several other ways to cut down on evaporation of water or to add water. But the plastic sheeting remains the best method of all. A seven day minimum cure is best. Actually 28 day cure strengths are used in a lot of figures, but that is a long, long time to hold up a project. Most concrete mixtures will reach quite a high percentage of their 28 day strength in a single week. Forms should be left in place for the entire curing period, both as support for the not cured concrete and as retardants to moisture loss. If the weather is hot and dry, the forms should be sprayed with water daily to cut down on water loss (Fig. 4-15).

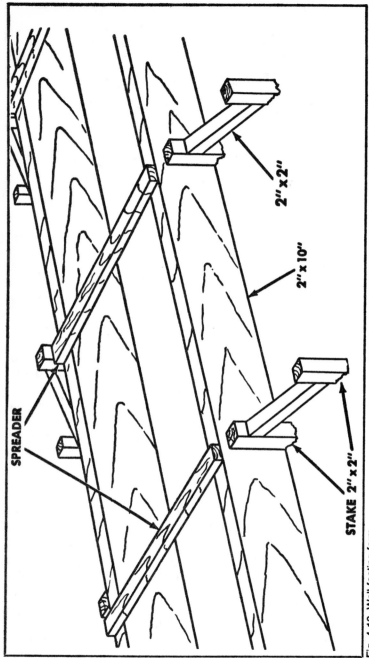

SPREADER

2" x 2"

2" x 10"

STAKE 2" x 2"

Fig. 4-10. Wall footing form.

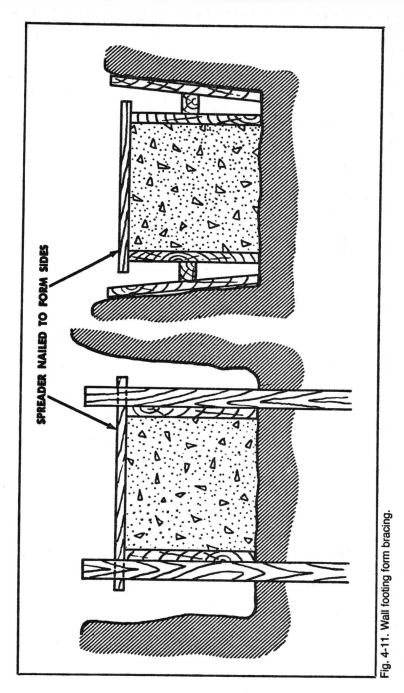

SPREADER NAILED TO FORM SIDES

Fig. 4-11. Wall footing form bracing.

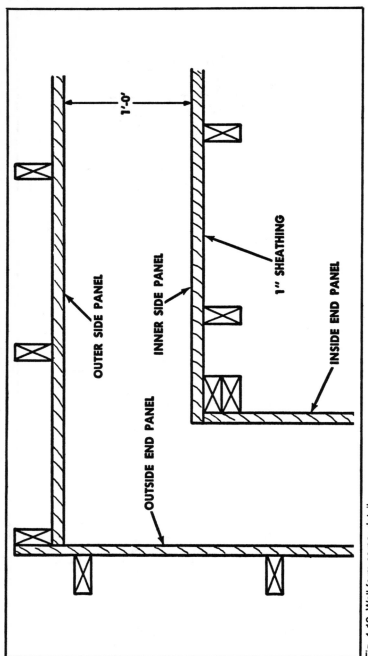

Fig. 4-12. Wall form corner details.

OUTER SIDE PANEL

INNER SIDE PANEL

1" SHEATHING

INSIDE END PANEL

OUTSIDE END PANEL

1'-0"

Table 4-9. Curing Methods.

Method	Advantage	Disadvantage
Sprinkling with water or covering with wet burlap.	Excellent results if constantly kept wet.	Likelihood of drying between sprinklings. Difficult on vertical walls.
Straw	Insulator in winter	Can dry out, blow away, or burn.
Moist earth	Cheap, but messy	Stains concrete. Can dry out. Removal problem.
Ponding on flat surfaces	Excellent results, maintains uniform temperature.	Requires considerable labor, undesirable in freezing weather.
Curing compounds	Easy to apply. Inexpensive	Sprayer needed. Inadequate coverage allows drying out. Film can be broken or tracked off before curing is completed. Unless pigmented, can allow concrete to get too hot.
Waterproof paper	Excellent protection, prevents drying	Heavy cost can be excessive. Must be kept in rolls, storage and handling problem.
Plastic film	Absolutely watertight, excellent protection. Light and easy to handle.	Should be pigmented for heat protection. Requires reasonable care and tears must be patched. Must be weighed down to prevent blowing away.

MORTAR

Without good mortar, no amount of care and work will give you a strong attractive and durable brick or block wall. Good mortar provides a strong bond between the masonry units and keeps excessive water from seeping into the joints and causing cracks and other problems in the structure. Bond strength is affected by several factors including the type and quantity of the cement used, the surface texture of the mortar bedding areas on the masonry units—the smoother the units the less surface area to which the mortar can bond—the plasticity of the mortar, and the water retention of the mortar. Naturally, craftsmanship in laying the brick is also of great importance in forming a good bond.

Because most masonry walls suffer their worst water leakage at the mortar joints, good mortar and good joint tooling are essential to the overall process of building a water resistant wall.

Mortar in proper form is plastic enough to be worked with a trowel without segregating. Segregating is caused by too much sand in the mixture. It is essential that only good, clean sand be used. Some masons insist on using what is known as sharp sand. But smooth sand forms just about as good as a bond, so that no extra searching time really needs to be spent if sharp sand is hard to come by in your locale. The proper plasticity, or workability, is obtained through the use of top quality materials in the proper proportions. Enough water is added to provide the correct consistency, but not so much as to give a slushy mixture that will run out of the joints and down the face of the brick.

Concrete block is not wetted before the application of the mortar, but certain types of brick should be. When the bricks are wetted, a good stream is directed on them until the water runs off the surfaces. The bricks are then allowed to surface dry and are bedded in mortar.

Two types of cement are used to produce mortar. Masonry cement can often be a bit hard to find, while portland cement is available everywhere I've been in this country. No matter what type of cement used, the durability of the mortar is controlled by the proportions of the materials. Mortar for ordinary service uses 1 measure of either masonry or portland cement. With portland cement, include one-half to 1 ¼ measures of hydrated lime and 4 ½ to 6 measures of masonry sand in a damp, loose condition. The masonry cement requires no lime, and needs only 2 ¼ to 3 measures of sand (all measures by volume). For masonry joints where service is apt to be severe, such as places with violent winds or

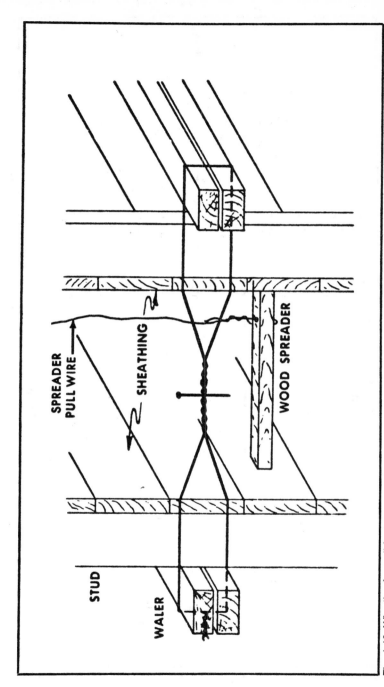

Fig. 4-13. Wire ties for wall forms.

SPREADER
PULL WIRE

SHEATHING

WOOD SPREADER

STUD

WALER

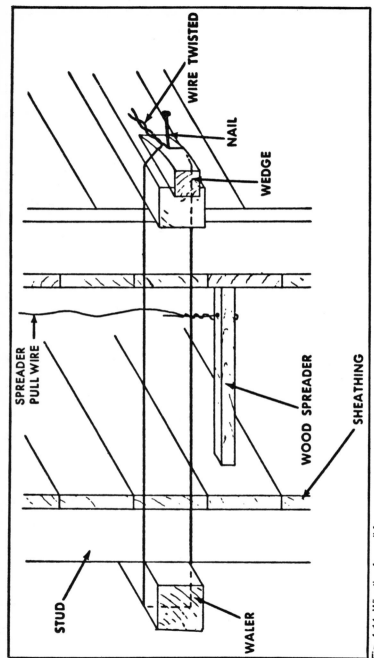

Fig. 4-14. Wire ties for wall forms.

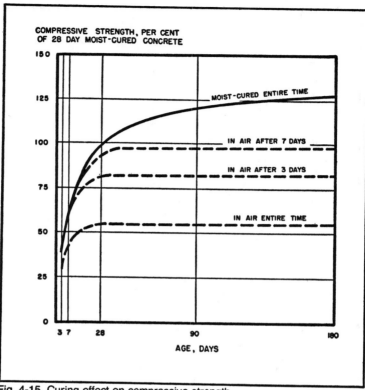

Fig. 4-15. Curing effect on compressive strength.

severe frost action, use 1 part of masonry cement, 1 part of portland cement and 4 ½ to 6 parts of sand. Or use 1 part of portland cement to about a one-quarter measure of hydrated lime and no more than 3 parts of sand.

Mortar types are designated as type M, type S, type N and type O. While all of these are seldom needed for around the home use, there is no harm in adding the needed specifications just in case.

Type M mortar uses 1 part portland cement, a quarter part of hydrated lime or lime putty and 3 parts of sand. It is suitable for general use, while also being recommended for below grade work and work in contact with the ground. It is especially good for retaining walls.

Type S mortar requires 1 part Portland cement, half a part of hydrated lime or lime putty and 4 ½ parts of sand. Type S is suited to general use and is specifically good where you need high resistance

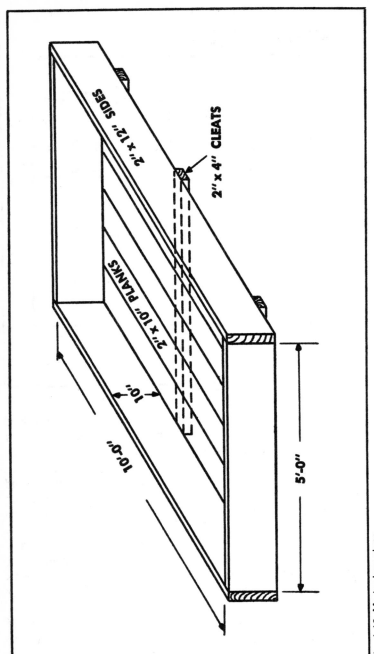

2" x 12" SIDES

2" x 4" CLEATS

2" x 10" PLANKS

10"

10'-0"

5'-0"

Fig. 4-16. Mortar box plans.

to lateral forces—such as where a wall is liable to be subject to high winds.

Type N mortar take 1 part portland cement, 1 part hydrated lime or lime putty and 6 parts of sand. This type is good for exterior use where weather conditions are liable to be nasty and where wind blown water might contain heavy concentrations of salt, as along the Atlantic coast.

Type O mortar involves 1 part portland cement, 2 parts of hydrated lime or lime putty, and 9 parts of sand. In one sense, this is a light duty mortar, but it is generally recommended for construction of load bearing walls where pressures do not pass 10 pounds per square inch. It is not strongly resistant to the action of moisture and of freezing.

Mortar is almost always mixed in much smaller amounts than is concrete. A mortar box can be bought, built or a wheel barrow can be used. All dry ingredients should be carefully mixed before adding water to ensure a uniform mixture. I find an ordinary garden hoe does a pretty good job in most cases, though the angles and corners in some mortar boxes and wheelbarrows might force you to do at least part of the mixing with a spade (Figs. 4-16 and 4-17).

Mortar that has stiffened a bit because of evaporation can sometimes be retempered. A small amount of extra water is added and the mortar is thoroughly remixed. It is extremely hard to tell the

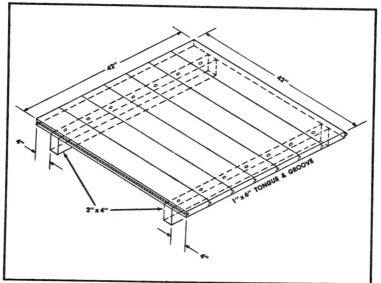

Fig. 4-17. Mortar board plans.

difference between initial setting and stiffening from evaporation. A rule of thumb can be used to keep from using partially set mortar, which would ruin the bonds in your joints. If the temperature is 80 degrees F or higher and the mortar was mixed much over two hours earlier, discard it instead of trying to retemper the material. If the mortar was first mixed more than three hours earlier on a day when the temperature doesn't quite hit 80 degrees, toss it out. It always seems that when you are doing this sort of heavy work that it is at least 80 degrees—in the shade which is usually 500 yards away.

Because the bond formed by the mortar joint is of such great importance in brick masonry, the few dollars, you might save by retempering mortar generally just isn't worth the chance of ruining the strength and durability of your work.

5

Ladders & Scaffolds

A great variety of ladders and a profusion of scaffolding types are available for those times when you need to work off the ground. Safety is always of paramount importance when working with these tools, as the sudden stop at the end of a fall from either can result in rather severe damage to the human carcass. Having stepped off, having slipped off, having ladders slip off the building with me on them, and once doing a rather intricate dance to stay on a two-story scaffold, I now operate under the theory that it is far, far better to be chicken than to put up with the aches and bruises that can result from even the shortest such fall.

HOOKS AND BOARDS

In general, for safe off the ground work, ladders are not the best for masonry work. You end up able to cart up only a few pieces at a time and the working space is extremely limited. Of course, two ladders with a scaffolding board hung on special hooks between the ladders can be used, but such makeshift scaffolding has an extreme lack of safety features even though the hooks and extendable boards are sold commercially by many companies. Either ladder could slip if both are not perfectly and firmly positioned. Also, there is almost no way you can provide yourself with a guard rail to prevent stepping off the scaffold board. Believe me, thinking you will concentrate on what you're doing and just not make that misstep doesn't work as well as you might hope. In the long run, you'll end up losing concentration on the job at hand and sooner or later make a mistake. Ladders are basically tools to allow you to get to higher spots. They are not tools for actually doing much work in or around those spots. Use your ladder to climb a scaffold, to reach a roof or to climb to the top of a wall and you'll improve your safety.

A single section portable ladder is about the simplest kind to be found today. Generally 30 feet or shorter in length, these can be very handy around most houses. They are also quite a bit lower in cost and weight than extension ladders. Extension ladders do provide you with extra height, but they also have added expense and weight. The extra weight makes longer ladders increasingly difficult to handle. Generally the longest readily available ladders are about 40 feet, which gives you about a 36-foot working height. This is only some 6 feet more than a single unit ladder. Of course, the 20-foot sections are much more easily transported.

WORKING HEIGHT

When you think of the working height of ladders, remember that the ladder will not be used at its maximum height. You cannot safely climb a 30-foot ladder that is standing straight up against the side of a building. Most safety experts explain that at least one-fourth of the length of the ladder is the distance you should place the base of the ladder from the wall (Fig. 5-1). If the ladder is held, or otherwise securely fastened, the distance can be lessened. You will need at least 16 inches of clearance at the back (wall side) of the ladder. After all, your toes have to go somewhere!

Ladder materials can be another source of contention. My preference for extension ladders is for fiberglass, but those are extremely expensive for homeowner use. Still, the weight is rela-

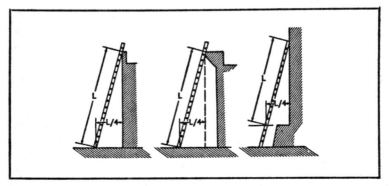

Fig. 5-1. Correct angles for placing a ladder.

tively low and the danger of electrocution should the ladder slap a power line while being set up or moved around is very low. Ladders made of wood are exceptionally heavy and subject to more rapid deterioration than either aluminum or fiberglass if care is not taken, and especially if they are stored outdoors. Aluminum ladders are lightweight, but they do provide one major danger. Contact with a power line would be a shock.

There is a variety of stepladders and speciality ladders. Some of these are exceptionally handy for such jobs as pointing brick walls, where little material and few tools are really needed. They are readily and easily moved from one spot to another.

BASIC SAFETY

Basic ladder safety involves regular inspection of your ladder. Even metal and fiberglass can wear out or get weathered to too great a degree. Wooden ladders should never be painted. Instead they should get a good coating of clear shellac or urethane varnish. Paint tends to hide flaws in the wood. At the job site, doors which could open and bump a ladder should be locked or blocked. Ladders must be placed so that the footing is secure and the top is on a good, solid surface. Do not place the ladder against a window pane or sash. Avoid using a ladder during a storm or during high winds. If you must place a ladder across a window frame, nail a 1 × 6 across the tops of a rail and at least a foot wider than the window opening.

Don't fling the ladder around when taking it down or moving it. Dropping the ladder on its side a couple of times is going to significantly cut down on its strength. Watch out for power lines. It is really best not to use metal ladders around power lines at any time. I'm reasonably sure that's the reason I've noted telephone repair and installation trucks in several states now carry fiberglass ladders. Wooden or fiberglass slung too hard into a poorly secured power line could drop the line onto you.

Ascend and descend a ladder facing the rungs and with the free use of both hands. Use a rope to lift any needed materials or tools after you get where you're going. Make sure the soles of your shoes aren't oily or greasy. I got a good scare one time after stepping into some spilled oil-base paint, without noticing. About half way up, I was dangling by one hand and the toes on one foot were trying to curl right through my boots to get a grip on the rung.

TEMPORARY PLATFORMS

Scaffolds are temporary platforms set up to provide a base for the working person and any needed tools or materials (Figs. 5-2 and

5-3). Naturally, the further you go from the ground, the higher the scaffold needs to be. Up until a decade or so ago, most scaffolding was built on the job and then torn down later. Now, metal scaffolding in take-apart-assemblies is, to a great extent, used instead of wooden structures. Still, for most people, the wooden scaffold probably offers the greatest combination of economy and flexibility. The metal types are fairly expensive and not available everywhere for rent. If they are easily rented in your locale, you'll probably at least save time and effort using them. And you might even save money.

Wood scaffolding, after it is taken down, can often be used as material for another project. However, there is going to be some loss of material as it is knocked apart and the nails pulled. Rough

Fig. 5-2. Professional brickmasons using patterned brick.

Fig. 5-3. Metal scaffolding in use for professional brick masons.

lumber can be used for scaffolding, and you might later use the lumber for the basis of a garden shed of some kind.

The scaffold will need to support a fair amount of weight in people, materials and tools. With masonry work the materials, if not the tools, weigh an awful lot. It must be strudy and it must be solidly nailed.

TRESTLE SCAFFOLDS

Trestle scaffolds are really not much more than boards laid across the tops of sawhorses and nailed to the sawhorse cross pieces. While the drawing here shows all wood construction (Fig. 5-4), Montgomery Ward sells a type of sawhorse leg that clamps a 2 × 4 or 2 × 6 very strongly. It will support quite a lot of weight. Each set is said to hold as much as 1500 pounds. They can be obtained in 24-inch and 30-inch lengths. I have three of the taller models, as

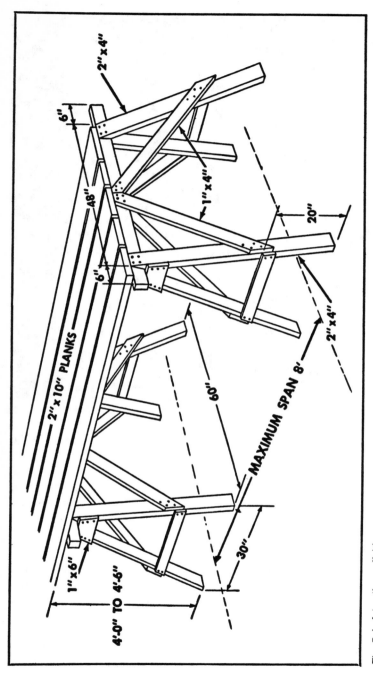

6"

2"x4"

1"x4"

20"

48"

6"

2"x4"

2"x10" PLANKS

60"

MAXIMUM SPAN 8'

1"x6"

30"

4'-0" TO 4'-6"

Fig. 5-4. A trestle scaffold.

well as several homemade sawhorses and others made with Stanley brackets. The homemade all-wood types are the most unwieldy and they take a while to build. For this type of scaffold, use 2 × 10s and keep the span down to 8 feet or less. Usually, these lower scaffolds are best for holding brick, mortar and tools at or near the height essential to keep you from having to continually bend down to pick them up when the wall gets above a certain height. Placing a mat under your knees and working on your knees is easiest when the wall is lower. Otherwise, you'll probably find yourself unable to straighten up at the day's end.

This type of scaffold is among the most dangerous for a very simple reason. Because they are seldom built much over 4 feet high, most people don't bother to add guard rails. This makes it far too simple to step off or slip off the edge. A 4-foot drop might not be much, but it can often cause painful and serious injury. If you are actually working on such a scaffold, I would recommend that you take that extra 6 inches of space at the back of the trestle and nail at least a 2 × 4 to it, with nails going into the flooring of the scaffold and into the trestle board. Then make a cross rail, or even two, nailing the cross rail to the inside of the uprights so that it cannot easily be knocked loose should you bump it hard. Do the same on each side.

The foot scaffold shown in Fig. 5-5 should never be used at heights over about 2 feet, preferably less. It is a time and back saver, but it can provide a severly twisted ankle or other body part.

POLE AND PUTLOG SCAFFOLDS

Pole and *putlog* scaffolds offer advantages in several ways over the simpler forms. This is particularly true when working with masonry where a scaffold should be able to support a load of 75 pounds per square foot. This is a truly heavy duty scaffold and the materials requirements reflect that. Poles will be no longer than 24 feet and at least of nominal 2 × 6 lumber. To go higher, you would have to use 4 × 4s or doubled 2 × 4s. The putlogs are doubled 2 × 4s or 2 × 8s laid on edge. Ledgers are 2 × 8s and braces are 1 × 6. Planking is of 2 × 10 material and guardrails are 2 × 4s.

These requirements are for a single pole scaffold, as shown in Fig. 5-6. As you can see, single pole does not mean there is only one pole. The front edge of the scaffold is supported by the building. Double pole, independent standing scaffolds are often needed for work where there is no back-up wall.

In building a scaffold, putlogs must be long enough to extend or overlap the poles at least three inches. They are face nailed to the

Fig. 5-5. A foot scaffold.

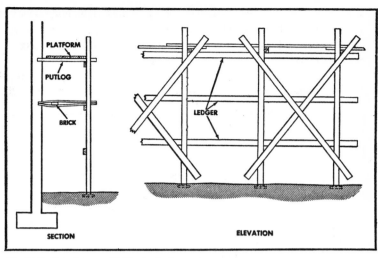

Fig. 5-6. A putlog scaffold, single leg.

pole and toe nailed to the ledger. Bearing blocks must be at least a foot long, of 2 × 6 lumber and notched to accept the end of the putlog. Each plank must be long enough to extend over three putlogs. All guard rails are nailed to the insides of the posts and all scaffolds must have guardrails (Fig. 5-7).

A few basic safety rules can help people safely on or around scaffolds. Any time one person is working above another, the person below must wear a hard hat for protection. A fixed ladder is used to get on and off the scaffold, not simply one leaned in place. Carting stuff up and down a non-secured ladder can be more hazardous than on a fixed ladder since a loose ladder is easier to knock out of position.

Don't work on a scaffold while it is raining. This would disrupt your masonry work anyway. Also, don't work during high winds or when conditions are icy. Tool containers should be lashed to the poles and tools not in immediate use should be placed in the containers to keep the tools from being knocked off the scaffold. Bring up only as much material as needed for a reasonable amount of working time. This will prevent overloading and tripping hazards. Never overload any scaffold.

BLOCK AND TACKLE RIGS

Don't toss materials and tools up or down from the scaffold. Many people do it from time to time, but the practice can be dangerous. Hand lines for lighter tools and materials, and block and tackle rigs for the heavier items are best. Such devices can save a lot of wear and tear on your back, since you won't need to cart a dozen or so bricks at a time up the ladder.

What the U.S. Army calls a gun tackle, one using two single sheave blocks, will provide enough mechanical leverage for the

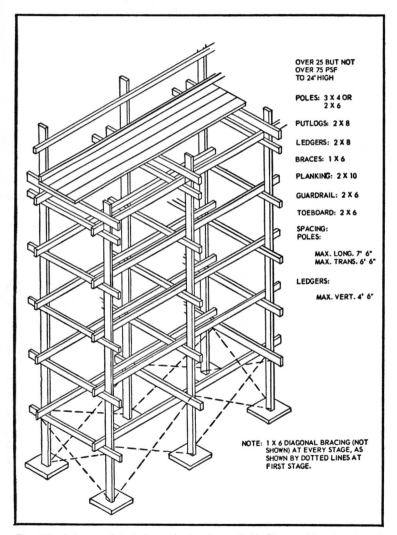

OVER 25 BUT NOT
OVER 75 PSF
TO 24' HIGH

POLES: 3 X 4 OR
2 X 6

PUTLOGS: 2 X 8

LEDGERS: 2 X 8

BRACES: 1 X 6

PLANKING: 2 X 10

GUARDRAIL: 2 X 6

TOEBOARD: 2 X 6

SPACING:
POLES:

MAX. LONG. 7' 6"
MAX. TRANS. 6' 6"

LEDGERS:

MAX. VERT. 4' 6"

NOTE: 1 X 6 DIAGONAL BRACING (NOT SHOWN) AT EVERY STAGE, AS SHOWN BY DOTTED LINES AT FIRST STAGE.

Fig. 5-7. A heavy duty independent pole scaffold. Diagonal bracing is not shown.

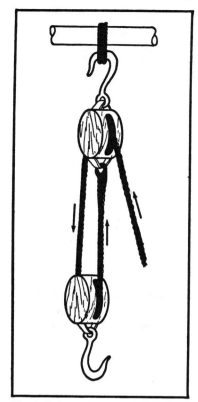

Fig. 5-8. Gun tackle.

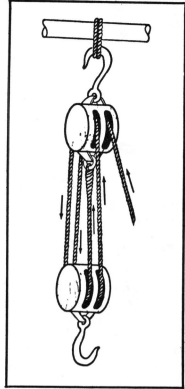

Fig. 5-9. Twofold purchase.

average person to haul up enough mortar and brick for a few hours work in two or three loads—the mechanical advantage is two (Fig. 5-8). For those with more help, and more extensive jobs, you might consider using a twofold purchase. This consists of two double blocks, giving a mechanical advantage of four, which should allow most anyone of normal strength to fairly easily haul up 250 pounds or more at a time (Fig. 5-9). Fancy gin poles and such are seldom needed, but it is a good idea to make sure the upper block is well supported.

Concrete Blocks

This is a book on brick masonry and I seem to sometimes drift a bit far afield with chapters on concrete and on concrete blocks. But much of the time, brick masonry involves back-up walls of other types of masonry. I feel like it makes sense to at least lightly cover such things just in case you're working on new construction and need to put in footings or a concrete block wall.

BLOCK SIZE

The first obvious difference between concrete block and brick as masonry units is size. While many block sizes are available, the nominal standard size is 8 × 8 × 16 inches. The second difference is quickly noted when you lift the block. A hollow core, load bearing block in the above size will weigh no less than 40 pounds and might even reach 50 pounds. The third difference is appearance. Concrete blocks do not make the most attractive final finish for construction work. For this reason, much work is covered with brick veneer or other types of veneers.

Figure 6-1 illustrates some typical sizes and types of concrete blocks. The variety is needed because of the many shapes and forms people like to build. It is best, whenever possible, to use a specific block designed for a specific purpose, rather than cutting a block to get the desired size or shape. This assumes modular planning on your part so that it is possible to determine the number of half blocks and lintels needed. Table 6-1 provides you with

Table 6-1. Nominal Length of Concrete Block Walls by Stretchers.

No. of stretchers	Nominal length of concrete masonry walls	
	Units 15⅝″ long and half units 7⅝″ long with ⅜″ thick head joints.	Units 11⅝″ long and half units 5⅝″ long with ⅜″ thick head joints.
1	1′ 4″	1′ 0″
1½	2′ 0″	1′ 6″
2	2′ 8″	2′ 0″
2½	3′ 4″	2′ 6″
3	4′ 0″	3′ 0″
3½	4′ 8″	3′ 6″
4	5′ 4″	4′ 0″
4½	6′ 0″	4′ 6″
5	6′ 8″	5′ 0″
5½	7′ 4″	5′ 6″
6	8′ 0″	6′ 0″
6½	8′ 8″	6′ 6″
7	9′ 4″	7′ 0″
7½	10′ 0″	7′ 6″
8	10′ 8″	8′ 0″
8½	11′ 4″	8′ 6″
9	12′ 0″	9′ 0″
9½	12′ 8″	9′ 6″
10	13′ 4″	10′ 0″
10½	14′ 0″	10′ 6″
11	14′ 8″	11′ 0″
11½	15′ 4″	11′ 6″
12	16′ 0″	12′ 0″
12½	16′ 8″	12′ 6″
13	17′ 4″	13′ 0″
13½	18′ 0″	13′ 6″
14	18′ 8″	14′ 0″
14½	19′ 4″	14′ 6″
15	20′ 0″	15′ 0″
20	26′ 8″	20′ 0″

(Actual length of wall is measured from outside edge to outside edge of units and is equal to the nominal length minus ⅜″ (one mortar joint).

nominal lengths for masonry walls using two different modular sizes, while Table 6-2 provides wall heights for both nominal 8-inch and nominal four-inch high masonry units.

FOOTINGS

Footings should be at least twice as wide as the wall resting on them (Fig. 6-2) and even wider if the ground is of the type that won't easily bear loads. As always, footings should be placed below the frost line and drainiage provided in wet areas. Many people suffer from wet basements. Mine now sometimes resembles a stream

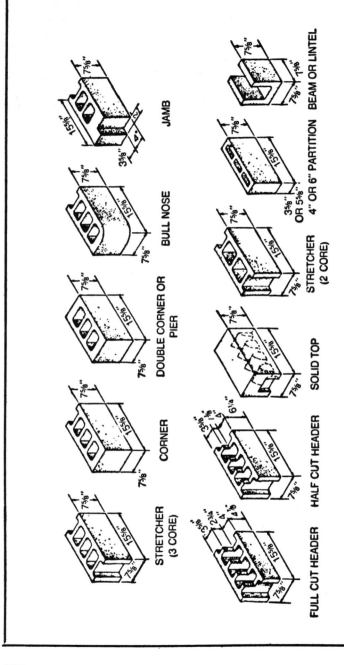

JAMB

7⅝"
15⅝"
3⅝"
4"

BULL NOSE

7⅝"
15⅝"
7⅝"

DOUBLE CORNER OR PIER

7⅝"
15⅝"
7⅝"

CORNER

7⅝"
15⅝"
7⅝"

STRETCHER (3 CORE)

7⅝"
15⅝"
7⅝"

BEAM OR LINTEL

7⅝"
7⅝"
7⅝"

4" OR 6" PARTITION

7⅝"
15⅝"
3⅝" OR 5⅝"

STRETCHER (2 CORE)

7⅝"
15⅝"
7⅝"

SOLID TOP

7⅝"
15⅝"
7⅝"

HALF CUT HEADER

3⅝"
1⅜"
6¼"
15⅝"
7⅝"

FULL CUT HEADER

3⅝"
2¾"
1⅜"
15⅝"
7⅝"

106

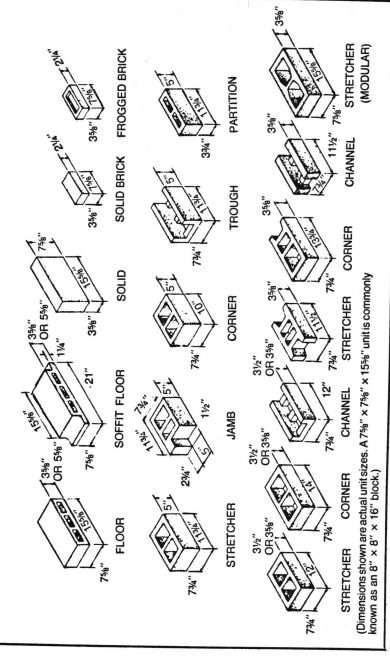

Fig. 6-1. Some typical sizes and shapes of concrete block.
(Dimensions shown are actual unit sizes. A 7⅝" × 7⅝" × 15⅝" unit is commonly known as an 8" × 8" × 16" block.)

107

during spring rains and the fix is going to be a lot more difficult and expensive than would have the original installation of Orangeburg perforated pipe laid in a bed of gravel along the front of the footing. This procedure will require the use of a back hoe and at least two days of digging.

Concrete block is laid much like brick (Fig. 6-3), with the exception of running mortar beads on the block in two lines outside the block cores. Butter the ends in the same manner (Fig. 6-4). With block, the units are large enough to check with a level as each one or two block are laid (Figs. 6-5 and 6-6). This is a lot more practical than the same job is for brick, but a guide line should still be used (Figs. 6-7, 6-8 and 6-9). Also, a story pole (Fig. 6-10) marked in the nominal increments of the units used (for height) will tell you quickly if your mortar joints, usually three-eights of an inch, are too large or too small. See Tables 6-1 and 6-2.

TOOLING JOINTS

Tooling joints is done after a section of wall is up. However, it is wise for those of us not working block or brick everyday to keep a

Table 6-2. Nominal Height of Concrete Block Walls by Courses.

No. of courses	Nominal height of concrete masonry walls	
	Units 7⅝" high and ⅜" thick bed joint	Units 3⅝" high and ⅜" thick bed joint
1	8"	4"
2	1' 4"	8"
3	2' 0"	1' 0"
4	2' 8"	1' 4"
5	3' 4"	1' 8"
6	4' 0"	2' 0"
7	4' 8"	2' 4"
8	5' 4"	2' 8"
9	6' 0"	3' 0"
10	6' 8"	3' 4"
15	10' 0"	5' 0"
20	13' 4"	6' 8"
25	16' 8"	8' 4"
30	20' 0"	10' 0"
35	23' 4"	11' 8"
40	26' 8"	13' 4"
45	30' 0"	15' 0"
50	33' 4"	16' 8"

(For concrete masonry units 7⅝" and 3⅝" in height laid with ⅜" mortar joints. Height is measured from center to center of mortar joints.)

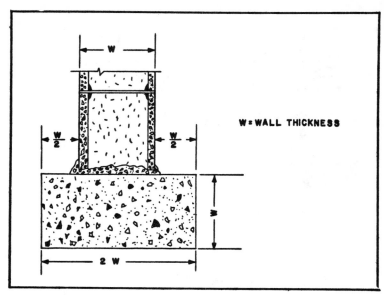

Fig. 6-2. Dimensions for masonry wall footings.

check on mortar setting and to do the joint tooling as soon as the first portion of the wall laid has mortar that is thumbprint hard. If the mortar will barely take a thumbprint, it is time to tool the joint. For concrete block, most experts recommend tooling the horizontal

Fig. 6-3. The mortar bed is laid with a furrow just as it is with brick.

Fig. 6-4. Only the outside lips of the concrete block are buttered for the vertical joints.

Fig. 6-5. The corner block will have a flat end and it is positioned first.

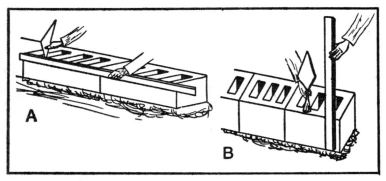

Fig. 6-6. You will get the best results if you level the courses after every four blocks are laid. Use the handle of the trowel to tap down any blocks that are too high.

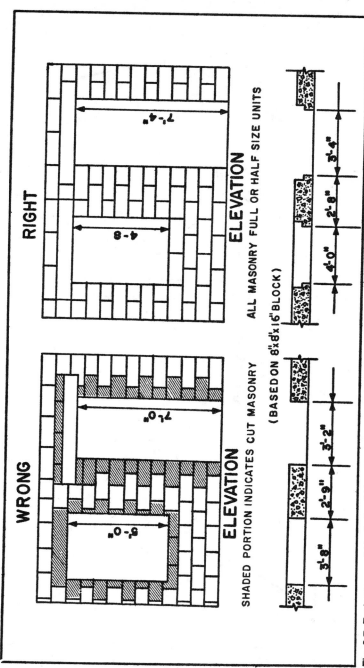

Fig. 6-7. Each course of block should be plumbed.

111

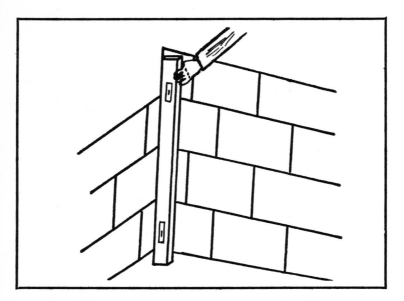

Fig. 6-8. Continue the check for plumb all the way up the wall.

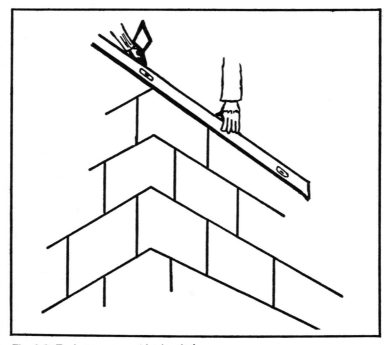

Fig. 6-9. Each course must be leveled.

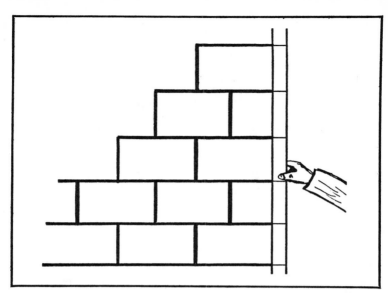

Fig. 6-10. Using a story pole, marked for the nominal height of the block being used, will aid in keeping mortar joints the same size.

joints first. Use only concave or V-joint shapes and make sure the jointing tool, as with brick, is a bit wider than the joint being worked.

CONTROL JOINTS

In long concrete block walls, control joints are needed. There is a chance of movement in the wall as time passes and control joints

Fig. 6-11. Excess mortar is cut from the joints with a trowel.

Fig. 6-12. Butter both ends of the closure block before insertion.

will permit that movement, within reason, and cut down on checking of the wall. Half blocks are used to form a continuous vertical joint, which is filled with mortar as are the other joints. As soon as the mortar sets up, or starts to, you rake it out to a depth of about three-quarters of an inch and then caulk the resulting gap with a good silicone material.

Precast lintels over doors are installed as the wall goes up, as are those over the windows, but the precast sills used under the windows are most often put in place after other construction is complete. These are all available in modular sizes designed to fit standard doors and windows.

From this point on, the laying of concrete block strongly resembles the laying of brick (Figs. 6-11 and 6-12). Whatever adaptations are needed will be rather obvious as you progress in the work and are not, in any case, extensive or particularly difficult.

7

Advanced Bricklaying

The cement sacks are in place, the sand is ready and the wheelbarrow is standing by. Your garden hose, bucket and hoe are all close at hand and you've made a three foot square mortar board and bought two trowels and a jointing tool. The yard seems overwhelmed by several thousand bricks and you're eager to get the projects rolling—learning, I hope, as you go.

But wait. Have you got the correct amounts of brick, sand, cement and lime! In most cases, the job isn't hard to figure. You will need to know how high and how long your wall or other structure will be. After that, decide what the thickness of your mortar joints will be and go to work to determine the amount of brick needed. If you underestimate on the mortar, sand and lime by even ˀn appreciable amount, it's no real chore to head for a building supply firm and pick up some more. But brick usually requires a wait for delivery, sometimes as much as two weeks if things are busy. That's a long delay up for any job.

Assume your wall is a single wythe one, using brick of 2¼-inch height. The wall is to be 5 feet high. You wish to use a three-eight inch mortar joint. A check of the chart in Table 7-1 shows that your wall will be 23 courses, (bricks) high, with an actual height of 5 feet and three-eighths an inch. As the charts in Tables 7-2 and 7-3 show, that's about as close as you can get to exactly 5 feet without adjusting a couple of mortar joints. This can be done, of course, but isn't really essential on exterior construction of a simple wall.

Table 7-1. Height of Courses Using 2¼ Inch Brick and ⅜ Inch Joints.

Courses	Height	Courses	Height
1	0' 2⅝"	51	11' 1⅞"
2	0' 5¼"	52	11' 4½"
3	0' 7⅞"	53	11' 7⅛"
4	0' 10½"	54	11' 9¾"
5	1' 1⅛"	55	12' 0⅜"
6	1' 3¾"	56	12' 3"
7	1' 6⅜"	57	12' 5⅝"
8	1' 9"	58	12' 8¼"
9	1' 11⅝"	59	12' 10⅞"
10	2' 2¼"	60	13' 1½"
11	2' 4⅞"	61	13' 4⅛"
12	2' 7½"	62	13' 6¾"
13	2' 10⅛"	63	13' 9⅜"
14	3' 0¾"	64	14' 0"
15	3' 3⅜"	65	14' 2⅝"
16	3' 6"	66	14' 5¼"
17	3' 8⅝"	67	14" 7⅞"
18	3' 11¼"	68	14' 10½"
19	4' 1⅞"	69	15' 1⅛"
20	4' 4½"	70	15' 3¾"
21	4' 7⅛"	71	15' 6⅜"
22	4' 9¾"	72	15' 9"
23	5' 0⅜"	73	15" 11⅝"
24	5' 3"	74	16' 2¼"
25	5' 5⅝"	75	16' 4⅞"
26	5' 8¼"	76	16' 7½"
27	5' 10⅞"	77	16' 10⅛"
28	6' 1½"	78	17' 0¾"
29	6' 4⅛"	79	17' 3⅜"
30	6' 6¾"	80	17' 6"
31	6' 9⅜"	81	17' 8⅝"
32	7' 0"	82	17' 11¼"
33	7' 2⅝"	83	18' 1⅞"
34	7' 5¼"	84	18' 4½"
35	7' 7⅞"	85	18' 7⅛"
36	7' 10½"	86	18' 9¾"
37	8' 1⅛"	87	19' 0⅜"
38	8' 3¾"	88	19' 3"
39	8' 6⅜"	89	19' 5⅝"
40	8' 9"	90	19' 8¼"
41	8' 11⅝"	91	19' 10⅞"
42	9' 2¼"	92	20' 1½"
43	9' 4⅞"	93	20' 4⅛"
44	9' 7½"	94	20' 6¾"
45	9' 10⅛"	95	20' 9⅜"
46	10' 0¾"	96	21' 0"
47	10' 3⅜"	97	21' 2⅝"
48	10' 6"	98	21' 5¼"
49	10' 8⅝"	99	21' 7⅞"
50	10' 11¼"	100	21' 10½"

Table 7-2. Height of Courses Using 2¼ Inch Brick and ½ Inch Joints.

Courses	Height	Courses	Height
1	0' 2¾"	51	11' 8¼"
2	0' 5½"	52	11' 11"
3	0' 8¼"	53	12' 1¾"
4	0' 11"	54	12' 4½"
5	1' 1¾"	55	12' 7¼"
6	1' 4½"	56	12' 10"
7	1' 7¼"	57	13' 0¾"
8	1' 10"	58	13' 3½"
9	2' 0¾"	59	13' 6¼"
10	2' 3½"	60	13' 9"
11	2' 6¼"	61	13' 11¾"
12	2' 9"	62	14' 2½"
13	2' 11¾"	63	14' 5¼"
14	3' 2½"	64	14' 8"
15	3' 5¼"	65	14' 10¾"
16	3' 8"	66	15' 1½"
17	3' 10¾"	67	15' 4¼"
18	4' 1½"	68	15' 7"
19	4' 4¼"	69	15' 9¾"
20	4' 7"	70	16' 0½"
21	4' 9¾"	71	16' 3¼"
22	5' 0½"	72	16' 6"
23	5' 3¼"	73	16' 8¾"
24	5' 6"	74	16' 11½"
25	5' 8¾"	75	17' 2¼"
26	5' 11½"	76	17' 5"
27	6' 2¼"	77	17' 7¾"
28	6' 5"	78	17' 10½"
29	6' 7¾"	79	18' 1¼"
30	6' 10½"	80	18' 4"
31	7' 1¼"	81	18' 6¾"
32	7' 4"	82	18' 9½"
33	7' 6¾"	83	19' 0¼"
34	7' 9½"	84	19' 3"
35	8' 0¼"	85	19' 5¾"
36	8' 3"	86	19' 8½"
37	8" 5¾"	87	19' 11¼"
38	8' 8½"	88	20' 2"
39	8' 11¼"	89	20' 4¾"
40	9' 2"	90	20' 7½"
41	9' 4¾"	91	20' 10¼"
42	9' 7½"	92	21' 1"
43	9' 10¼"	93	21' 3¾"
44	10' 1"	94	21' 6½"
45	10' 3¾"	95	21' 9¼"
46	10' 6½"	96	22' 0"
47	10' 9¼"	97	22' 2¾"
48	11' 0"	98	22' 5½"
49	11' 2¾"	99	22' 8¼"
50	11' 5½"	100	22' 11"

Length at 20 feet is broken down to 240 inches and divided by nominal brick length, such as 8 inches. Each course will be 30 bricks long and the wall will be 23 bricks high, for a total of 490 bricks in all. With a waste allowance of 10 percent, you would need about 540 bricks (Table 7-4).

Mortar estimates are more subject to variation since neat masons need less than sloppy masons and since there are so many different mortar mixes possible. At a rough guess, the above wall could take from two to three bags of portland cement, with sand and lime in proportion to that. Table 7-5 shows the approximate conversations of aggregates to weight, instead of volume, for those areas where it must be bought in such a manner.

At this point, I will assume that all footings and foundation walls are in place for your brick wall and step right into the work of laying out a common bond brick wall. First, lay out the brick along the foundation without using mortar (Fig. 7-1). You will want to make adjustments in head joint widths, if needed, so that the total length includes one half brick. In any masonry work using regular sized units, such as brick and concrete block, the corners are laid first. Therefore, if your wall turns a corner, this is where the first bed of mortar will be laid.

CORNERS

The job of laying corners is called laying leads. The first courses of a wall should first be laid dry. Then the leads serve as guides for the construction of the remainder of the wall. As I mentioned earlier, you should lay the mortar bed an inch deep and with a slight furrow or groove down its center to receive the first course. The face tier is laid on the corners and each will be built up for some 6 to 8 courses, or to the next header course, whichever comes first. Normally, the next header course will fall within the above range. But with common bond walls that are not heavily involved in structural support of anything beyond their own weight, the design of header courses is up to you. Usually, this first course in a common bond wall is a header course, but there is no rule that says it must be. If you'd prefer to have the second, third or even fourth course be the first header course, go right ahead.

Cut two, three-quarter closures before laying the mortar bed. Place the first one and butter the end of the other, butting it against the first. This forms your first head joint at the corner. This head joint should be one-half inch thick. Cut off the excess mortar squeezed out of the head joint and use either a plumb rule or level to

Table 7-3. Height of Courses Using 2¼ Inch Brick and ⅝ Inch Joints.

Courses	Height	Courses	Height
1	0' 2⅞"	51	12' 2⅝"
2	0' 5¾"	52	12' 5½"
3	0' 8⅝"	53	12' 8⅜"
4	0' 11½"	54	12' 11¼"
5	1' 2⅜"	55	13' 2⅛"
6	1' 5¼"	56	13' 5"
7	1' 8⅛"	57	13' 7⅛"
8	1' 11"	58	13' 10¾"
9	2' 1⅞"	59	14' 1⅝"
10	2' 4¾"	60	14' 4½"
11	2' 7⅝"	61	14' 7⅜"
12	2' 10½"	62	14' 10¼"
13	3' 1⅜"	63	15' 1⅛"
14	3' 4¼"	64	15' 4"
15	3' 7⅛"	65	15' 6⅞"
16	3' 10"	66	15' 9¾"
17	4' 0⅞"	67	16' 0⅝"
18	4' 3¾"	68	16' 3½"
19	4' 6⅝"	69	16' 6⅜"
20	4' 9½"	70	16' 9¼"
21	5' 0⅜"	71	17' 0⅛"
22	5' 3¼"	72	17' 3"
23	5' 6⅛"	73	17' 5⅞"
24	5' 9"	74	17' 8¾"
25	5' 11⅞"	75	17' 11⅝"
26	6' 2¾"	76	18' 2½"
27	6' 5⅝"	77	18' 5⅜"
28	6' 8½"	78	18' 8¼"
29	6' 11⅜"	79	18' 11⅛"
30	7' 2¼"	80	19' 2"
31	7' 5⅛"	81	19' 4⅞"
32	7' 8"	82	19' 7¾"
33	7' 10⅞"	83	19' 10⅝"
34	8' 1¾"	84	20' 1½"
35	8' 4⅝"	85	20' 4⅜"
36	8' 7½"	86	20' 7¼"
37	8' 10⅜"	87	20' 10⅛"
38	9' 1¼"	88	21' 1"
39	9' 4⅛"	89	21' 3⅞"
40	9' 7"	90	21' 6¾"
41	9' 9⅞"	91	21' 9⅝"
42	10' 0¾"	92	22' 0½"
43	10' 3⅝"	93	22' 3⅜"
44	10' 6½"	94	22' 6¼"
45	10' 9⅜"	95	22' 9⅛"
46	11' 0¼"	96	23' 0"
47	11' 3⅛"	97	23' 2⅞'
48	11' 6"	98	23' 5¾"
49	11' 8⅞"	99	23' 8⅝"
50	11' 11¾"	100	23' 11½"

Table 7-4. Quantities of Materials for Brick Walls.

Wall area sq ft	Wall thickness in inches							
	4 inches		8 inches		12 inches		16 inches	
	Number of bricks	Cu ft mortar	Number of bricks	Cu ft mortar	Number of bricks	Cu ft mortar	Number of bricks	Cu ft mortar
1	6.17	.08	12.33	.2	18.49	.32	24.65	.44
10	61.7	.8	123.3	2	184.9	3.2	246.5	4.4
100	617	8	1,233	20	1,849	32	2,465	44
200	1,234	16	2,466	40	3,698	64	4,930	88
300	1,851	24	3,699	60	5,547	96	7,395	132
400	2,468	32	4,932	80	7,396	128	9,860	176
500	3,085	40	6,165	100	9,245	160	12,325	220
600	3,712	48	7,398	120	11,094	192	14,790	264
700	4,319	56	8,631	140	12,943	224	17,253	308
800	4,936	64	9,864	160	14,792	256	19,720	352
900	5,553	72	10,970	180	16,641	288	22,185	396
1,000	6,170	80	12,330	200	18,490	320	24,650	440

*Quantities are based on ½-inch-thick mortar joint. For ⅜-inch-thick joint use 80 percent of these quantities. For ⅝-inch-thick joints use 120 percent.

check the level of these bricks. Make sure the face edges of the two bricks are flush with the foundation.

Butter the side of the next brick—brick *c* in Fig. 7-1—and lay it in place. Make sure its end is flush with the foundation. Next comes brick *d*. The levels of the four bricks in place are again checked. The level of most of the bricks laid in the corners will be checked as you go along, especially those in the lower couple of courses, since this is the basis for a good wall, with straight mortar joints. Misjudging the level at this point can call for some drastic fiddling with joints at a later point to return things to level. And this can also make the finished job look terrible. Once these four bricks are in place, the quarter closures—*e* and *f* in Fig. 7-1—are cut and placed according to earlier directions for laying closure bricks (Chapter 3). Level is again checked after excess mortar is removed.

Table 7-5. Aggregates by the Pound.

Class concrete (figures denote size of coarse aggregate in inches)	Estimated 28-day compressive strength, (pounds per square inch)	Cement factor, bags (94 pounds) of cement per cubic yard of concrete, freshly mixed	Maximum water per bag (94 pounds) of cement (gallons)	Fine aggregate range in percent of total aggregate by weight	Approximate weights of saturated surface-dry aggregates per bag (94 pounds) of cement	
					Fine aggregate (pounds)	Coarse aggregate (pounds)
(1)	(2)	(3)	(4)	(5)	(6)	(7)
B-1	1500	4.10	9.50	42-52	368	415
B-1.5	1500	3.80	9.50	38-48	376	498
B-2	1500	3.60	9.50	35-45	378	567
B-2.5	1500	3.50	9.50	33-43	373	609
B-3.5	1500	3.25	9.50	30-40	378	702
C-1	2000	4.45	8.75	41-51	329	387
C-1.5	2000	4.10	8.75	37-47	338	467
C-2	2000	3.90	8.75	34-44	338	529
C-2.5	2000	3.80	8.75	32-42	332	565
C-3.5	2000	3.55	8.75	29-39	334	648
D-0.5	2500	5.70	7.75	50-60	282	231
D-0.75	2500	5.30	7.75	45-55	288	288
D-1	2500	5.05	7.75	40-50	279	341
D-1.5	2500	4.65	7.75	36-46	287	413
D-2	2500	4.40	7.75	34-42	288	471
D-2.5	2500	4.25	7.75	32-40	287	509
D-3.5	2500	4.00	7.75	29-37	285	578
E-0.5	3000	6.50	6.75	50-58	238	203
E-0.75	3000	6.10	6.75	45-53	240	249
E-1	3000	5.80	6.75	40-48	233	297
E-1.5	3000	5.35	6.75	36-44	239	359
E-2	3000	5.05	6.75	33-41	241	410
E-2.5	3000	4.90	6.75	31-39	238	441
E-3.5	3000	4.60	6.75	28-36	237	503

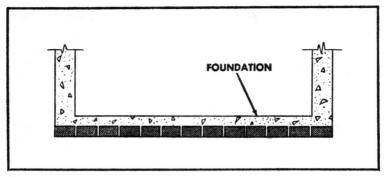

Fig. 7-1. Brick laid without mortar.

From this point, brick *g* is buttered along its face and laid, excess mortar cut off and the process repeated for the rest of the bricks along the first course in that corner. Run at least six bricks with a header course, and four if you're laying a stretcher course.

With the header course corner laid, spread another 1-inch bed of mortar along the tops of the bricks—after checking for level in both directions, as the dotted lines in Figure 7-6 show—and proceed to lay your first stretcher course. To lay a stretcher course as the first course, in this type of double wall, requires that both stretcher courses be laid before a header course can be put in above them. The outside stretcher course is laid first and the mortar is carefully applied Figs. 7-2A and 7-2B so that little excess slops over onto the part of the header course that will not yet be covered.

Full brick is used at the first corner on the stretcher course and the brick laying procedure carries on now until the wall reaches the height of the next header course. At that time, the inside stretcher course is laid and the new header course laid just as was the first. The wall is then continued in the same manner, with a careful check kept on the plumb of the corners (Fig. 7-3).

The joints are tooled in the manner you decide as soon as they get thumbprint hard. You can use a line, after the corners are in place, to help provide a guide to keeping the wall level along its length. This way you will not have to check every time one or two bricks are laid. A nail can be pushed into a joint at each corner and a mason's cord wrapped around two pieces of scrap brick and hung over the nails (Figure 7-4).

If you want to build a 12-inch instead of an 8-inch common bond wall, things are a bit more complex, but not really too much so. What you do, essentially is in your first course, lay both a header and

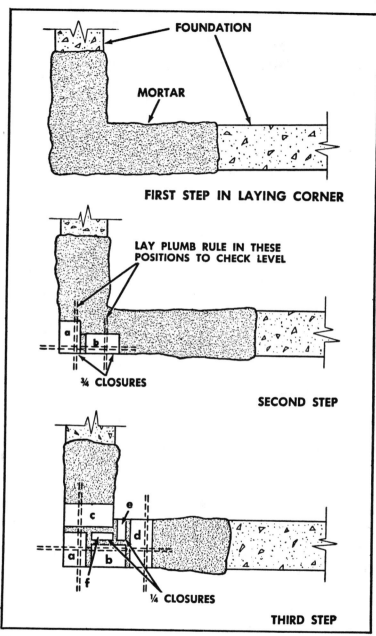

FOUNDATION

MORTAR

FIRST STEP IN LAYING CORNER

LAY PLUMB RULE IN THESE POSITIONS TO CHECK LEVEL

a

b

¾ CLOSURES

SECOND STEP

c

e

d

a

b

f

¼ CLOSURES

THIRD STEP

Fig. 7-2A. The first 3 steps in laying a corner lead for an eight-inch common bond brick wall.

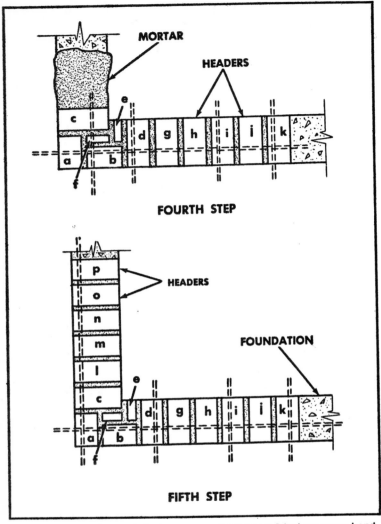

MORTAR

HEADERS

FOURTH STEP

HEADERS

FOUNDATION

FIFTH STEP

Fig. 7-2B. The final steps in laying a corner lead for an 8-inch common bond brick wall.

a stretcher course (Fig. 17-5). Once the first course is in, again using three-quarter closures at the corners, a second course with the header course can be laid to the inside. This assumes the first course of headers was to the outside, which is not a requirement—your needs and desires affect which goes where). (Fig. 7-6) The third course can consist of three stretcher courses or the wall can

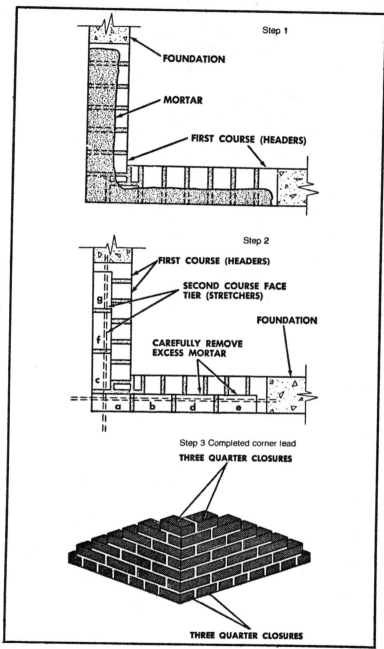

Fig. 7-3. The second course of a corner lead for an eight-inch brick wall.

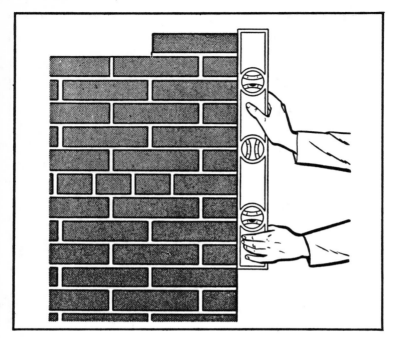

Fig. 7-4. Plumbing a corner.

now become a cavity wall. More on cavity wall construction a bit later (Fig. 7-7).

The top of any brick wall under construction will need protection at night. Using polyethelene sheeting is probably the most economical method of providing the protection.

OPENINGS

If your brick wall is to enclose a house, or a portion of another building, openings must be left for doors and windows as required. Normally, bricks are not cut to height to provide window openings. The top of one course becomes the base for the sill. The distance is easy enough to determine in courses if you have already planned the height of the window openings. Under normal circumstances, home construction keeps the height of the tops of windows and doors at the same level to provide a cleaner line.

The final sill course is a rowlock course, with the bricks set on edge and angled downward to provide water drainage from the window. This rowlock course is normally set to take up, at the back edge, the same amount of space as two regular courses would take.

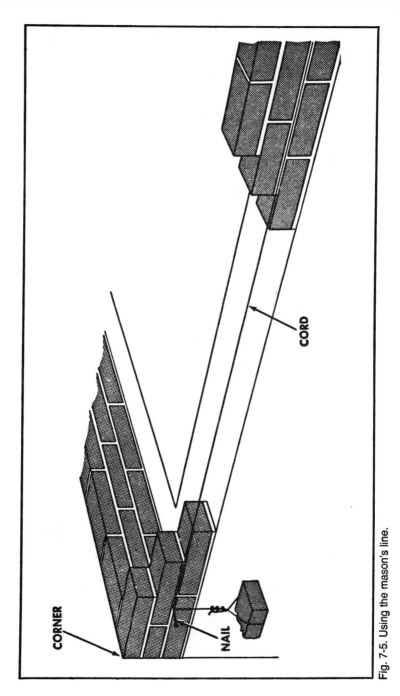

Fig. 7-5. Using the mason's line.

127

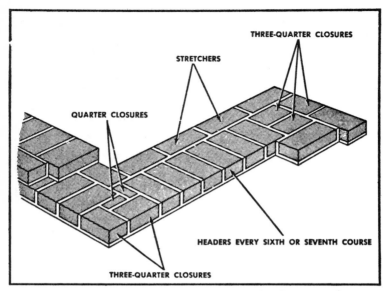

Fig. 7-6. The first course of a 12-inch common bond wall.

Joints in window sill rowlock courses must be carefully filled and carefully tooled to ensure a great degree of (Fig. 7-8) water tightness.

Once the rowlock course is in place, the window can be set on it. Temporary bracing will be needed for several days. Actually, the bracing isn't really needed as bracing once the courses are laid past about a third of the window height. Leaving it in place for several days allows the mortar in the courses around the window time to set properly. As the courses go up around the window, care should be

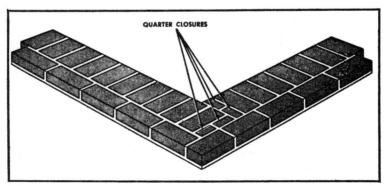

Fig. 7-7. The second course of 12-inch common bond wall.

128

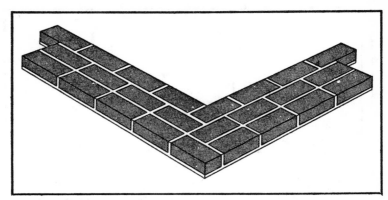

Fig. 7-8. The third course of a 12-inch common bond wall.

used. The top brick course, at the top of the window frame, should be no more than one-quarter of an inch above the top of the frame. In brick masonry with window openings, the brick is laid to the top of

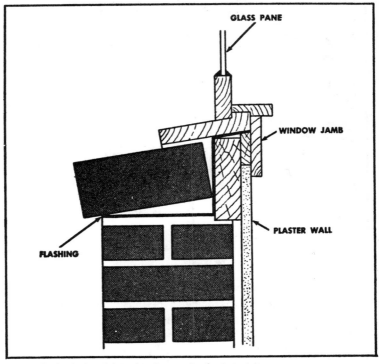

Fig. 7-9. A window joint, showing rowlock window course.

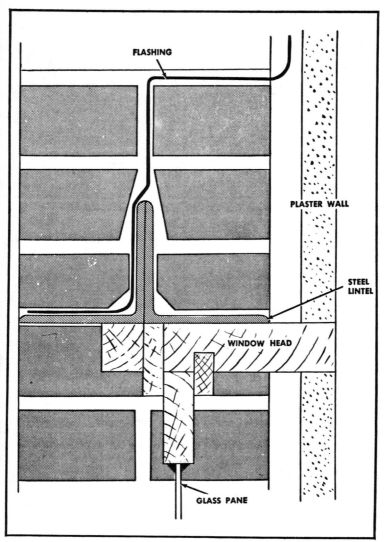

Fig. 7-10. Flashing at a window opening.

the window before continuing up the walls with the corner leads since it may be necessary to adjust the thickness of several of the mortar joints to get a final, correct fit on the top course. Naturally, this means that the mortar joint down the courses to the corner lead must also be adjusted. This can't be done until you know what the adjustment is and in which joints (Fig. 7-9).

Lintels are placed over windows and doors to carry the load of the wall above them. You need only one try at opening a window with either an improperly placed lintel or no such bracing to realize the true description of futility. The lintel ends rest on the final course of brick that is level or within a quarter inch of being so with the top of the window frame. The lintel ends are firmly embedded in mortar, using the same joint thickness as you have for the rest of the wall. Lintels are available, as Figs. 7-10 and 7-11 show, in precast concrete and steel. Steel lintels are preferred where you don't want the look of concrete anywhere on a wall. They can be hidden behind a mortar joint. Wood lintels are sometimes available, but tend to be a bit on the weak side if not perfectly made. These also add a different dimension to the looks of the window. If wood is used, it should always be the pressure treated kind such as Kopper Company's Outdoor brand.

Lintel size is determined by the span over the door or window frame, as well as by the wall thickness. Figure 7-12 shows some sizes. Precast concrete lintels come in the appropriate size for their width and depth. Please note that in no case should wood lintels be used for brick wall openings over three feet wide. The primary reason for this is retention of wall integrity in a fire. A wood lintel wider than 3 feet might burn through, allowing the wall above to collapse into a heap.

Door openings work in pretty much the same manner as do window openings, with a much lower starting point of course. Generally, pieces of wood cut to the size of a half closure brick are

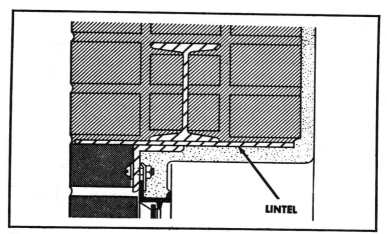

Fig. 7-11. A steel lintel.

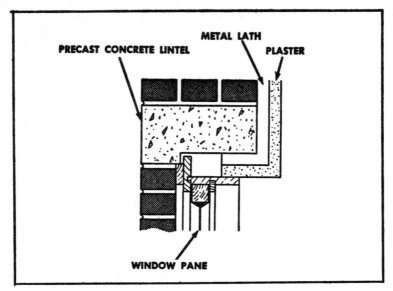

Fig. 7-12. A precast concrete lintel.

laid in mortar just as if they were brick in order to provide an anchor for the screws or nails used to hold the door frame in place. Again, wood used in such a way should be of the pressure treated variety. At least five such wood blocks should be used on each side of the door frame to provide secure anchorage and to allow for any needed levelling and plumbing of the door (Table 7-6).

CORBELING

Corbeling sounds fancy, looks fancy and isn't all that hard to do. Corbeling is nothing more than bricks set out above those in the course or courses below to form a self-supporting project. In essence, it is a form of cantilever on a small scale. Most often used in chimney construction, corbeling provides a bit of extra weather resistance. It can be used decoratively and it has other practical applications with just a few simple rules kept in mind.

First, corbeling is best done with header courses to allow greater bonding of the corbeled bricks. The first corbeled course can be a stretcher course if that is essential. Next, no single course below it. Also, the total distance the corbel projects should never be more than the total thickness of the original wall (Fig. 7-13).

For greatest strength, use even more care when corbeling to get carefully filled and tooled mortar joints so that the wall will retain

Table 7-6. Common Lintel Sizes.

Wall thickness	Span							
	3 feet		4 feet* steel angles	5 feet* steel angles	6 feet* steel angles	7 feet* steel angles	8 feet* steel angles
	Steel angles	Wood					
8"	2-3 × 3 × ¼	2 × 8 2-2 × 4	2-3 × 3 × ¼	2-3 × 3 × ¼	2-3½ × 3½ × ¼	2-3½ × 3½ × ¼	2-3½ × 3½ × ¼
12"	2-3 × 3 × ¼	2 × 12 2-2 × 6	2-3 × 3 × ¼	2-3½ × 3½ × ¼	2-3½ × 3½ × ¼	2-4 × 4 × ¼	2-4 × 4 × 4¼

*Wood lintels should not be used for spans over 3 feet since they burn out in case of fire and allow the brick to fall.

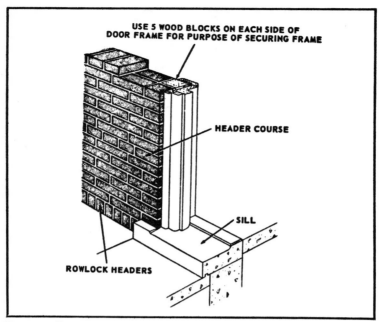

USE 5 WOOD BLOCKS ON EACH SIDE OF
DOOR FRAME FOR PURPOSE OF SECURING FRAME

HEADER COURSE

SILL

ROWLOCK HEADERS

Fig. 7-13. Construction at a door opening.

as much strength as possible. If you're building a wall that will support heavy loads, it would almost certainly be an excellent idea to consult an engineer and get the stresses figured to make sure the structure won't cave in.

BRICK ARCHES

Brick archwork looks exceedingly complicated and it can easily intimidate the novice brick layer. It should. An arch is not the place to start learning how to lay brick! But a properly constructed brick arch can support an extremely heavy load, because of its curved shape, though elliptical arches don't have all that much of a curve (Fig. 7-14). Mortar joint thickness is going to vary on each and every brick laid in an arch, obviously. The bottom of the arch brick will have a narrower joint than will the top, but in no case should you ever lay in an arch brick with less than one-quarter inch of mortar at the bottom. Great care has to be taken to be sure that each and every joint in the arch will be well filled with mortar.

A bit of carpenter's artwork is needed to provide a temporary support for brick archways under construction. Somewhere, there might be a mason who can put up a brick arch without such a support, but I've never met one and I expect to do so about two days after I meet Merlin the Magician. You will, I hope, have drawings indicating the correct shape and size of the arch you wish to construct. From these, make a tracing on a large sheet of brown paper. Check the dimensions carefully before cutting your template for the supports from the brown paper. Now, take two sheets of plywood (or one sheet if the arch is a small one) and using the pattern, lay out the arch shape. Cut two supports from the three-quarter inch plywood using a saber saw. Using eight or 10 penny nails, nail this to 2 × 4s placed between them. The resulting support surface will be about 3 inches wide and sufficient to support the bricks as the mortar sets, without allowing any twisting.

As Fig. 7-15 shows, the template is held in place on concrete blocks, with wood props to do the final holding. Wood is easily cut to the exact length and when the mortar has set, a couple of pops with

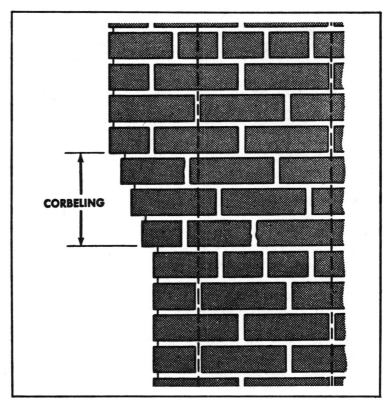

Fig. 7-14. Corebeled brick wall.

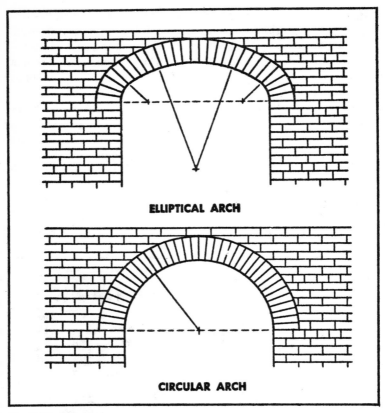

ELLIPTICAL ARCH

CIRCULAR ARCH

Fig. 7-15. Arch shapes.

a hammer on the wood props removes them and drops the template out of the way.

Arch construction begins at the two ends (abutments to be technical about it). The brick is laid from both ends, swapping back and forth each few bricks and working to the center. The center, or key, brick is the last to be laid. Arches are always laid out so that you do not need to cut a brick to fit. The spacing of the brick is determined quite simply by laying dry brick around the temporary support, on the ground. Spacing is adjusted until the key brick is dead center on the arch pattern. Then you take a pencil and mark the postion of each and every brick to go into the arch on both sides of the support board.

Because of the need for really good setting, I would leave the arch support in place for at least several days. In cases where the weather isn't very agreeable, it is better to allow a few extra days for the mortar to cure rather well than to get bopped on the head with falling bricks as soon as the props are knocked out from under the arch support.

WATERTIGHT WALLS

Even with good bonding, there is sometimes a bit less of water resistance in a masonry wall than you might care to find. In most cases this will occur with below ground masonry walls, only a part of which will be made up of brick. With brick, water passing through will almost always move through the mortar joints. With concrete block, the joints and sometimes the block itself can allow seepage. Head joints rather then bed joints are the most likely source of seepage. Again, the best solution is good bonding. Use bricks that are damp but not dripping wet and with well tooled joints. On above ground walls and other structures, the use of sealing compounds, paints changes the look of the brick. And I can assume that at least one of the reasons brick units are used in any structure is for their looks. Careful craftsmanship will help retain the brick look, as you can materially reduce the number of cracks in mortar joints that might allow the passage of water.

For double wythe walls where leakage might be a problem, there is a good solution that is called parging. In this method, the back of the bricks in the face tier are parged, or coated, with a troweled on mixture of very cement-rich mortar before you lay the bricks in the backing courses. The process of parging (also called

Fig. 7-16. Use of a template in arch construction.

back plastering) requires a coating of no less than three-eighth of an inch thickness. The mortar joints on the back of the face courses must be cut flush for the coating to do its job (Fig. 7-16).

For the partially below ground wall where water pressure can be severe, a combination of drainage tile, loose stone, and, possibly, a membrane between the two wythes of the wall, will do a good job of preventing leakage. The drain tile, if used, should be at least 4 inches in diameter and well perforated. Loose gravel can be used without the drain tile, though it isn't as efficient so use probably should be confined to areas where the water problem is relatively minor (Fig. 7-17).

COATINGS

Foundation walls, seldom made of brick these days, below ground level are usually well protected from water seepage if drain tile is laid in gravel along the footings. The wall is given two good coats of bituminous mastic along its full outside surface (all the surface that is to be below grade). Even asphalt will do the job rather well. Such coatings are applied with heavy broomlike brushes or mops.

Above the ground level walls are where waterproof coatings must be carefully selected for transparency and a lack of a tendency to flake or peel. A water solution of sodium silicate will often do the job in less extreme cases, while you might have to go to a urethane varnish or special brick sealing compound where seepage is worse. In cases where you are working with an old brick wall and don't care to worry about the retention of the brick appearance, then most good quality house paints will seal the wall—after it has been carefully cleaned. Portland cement paint can be applied to damp surfaces (use a garden hose set on fine spray) and a whitewash brush is used.

Before using any transparent or non-transparent coating on a brick wall, all cracks in the mortar joints should be repaired. The method, called pointing or tuckpointing, is simple. Carefully chip the mortar out of the cracked joint to a depth of 2 inches. Clean the resulting hole with water. Make sure it is really scrubbed clean of old pieces of mortar and brick dust. Use the type S mortar, described earlier, for tuckpointing. While the brick surface is still wet, a coating of thinned mortar (thinned to a consistency about like flour paste) is applied to the joint being repaired. Before the thinned coating sets, the type S mortar is applied and the resulting new joint is carefully tooled.

Fig. 7-17. Parging a wall.

FIRE BRICK

Fireplaces, barbecues and incinerators are items you might want to build. All require the use, in certain areas, of brick more resistant to heating and cooling than standard brick. Fire brick, made of fire clay, is essential to long life and good performance. It is not much different to work with than standard brick, with a couple of minor exceptions.

Fire brick is larger, with the most common size being 9 × 4½ × 2½ inches. There is a need for joints no thicker than one-quarter of an inch when laying fire brick. The thicker the joint the more likely it is to fail in constant use. Mortar used to lay fire brick is not one of the standard types. It must consist of fire clay mixed with water and it is mixed to the consistency of a thick cream of flour paste instead of the thicker consistency of standard mortars.

The brick is dipped in the mortar in a manner that will coat all the faces except the one to be exposed. The brick is then set in place. After a few courses have been laid, you can expect to have to tap them gently into place with a mason's hammer fire bricks are set as tightly as possible to prevent heat penetration to the outside structure. Damage might result from that heat penetration. As I said, the joints should be no more than one-quarter of an inch thick. Less is better, and the more tightly the bricks fit together, the better the final result. Fire brick is laid so that half of one brick overlaps the one below, with the head joints falling in the center of the bricks above and below them. In essence, it is a running bond style.

SPECIAL WALLS

The common bond walls already covered are solid brick walls. They are excellent for many purposes, from decoration to retaining

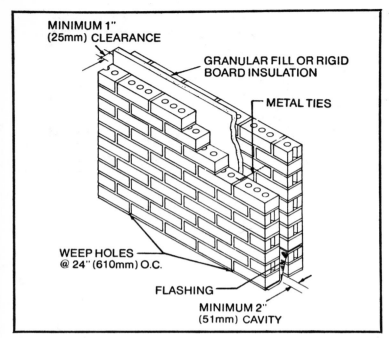

Fig. 7-18. Foundation drain details.

wall use. But for heat retention and less sound transfer, other types of walls are often preferred or needed. A solid brick wall would have to be something like 4 feet thick, and possibly more, to equal a standard frame walls thermal insulation value. Even that is considered inadequate with fuel oil becoming scarcer and more costly almost day by day. Many new homes are using different framing methods and heavier insulation to attain greater R (resistance to heat loss) values. A brick wall can be constructed to cut away down on heat loss.

There will still be heat losses, of course, as there always is in any type of building, but with modern insulations those losses can be cut to a minimum. The greatest heat loss occurs through the metal ties used to connect (for structural integrity) one wythe to another.

The most important special brick wall is the cavity wall. A properly constructed brick cavity wall is impervious to water penetration. While the exterior section of the wall might not be completely waterproof, whatever moisture does get through that outer wythe runs down the inside and is directed out at the wall's base by means of flashing and weep holes. All cavity walls must have flashing installed all along their bases, with weep holes every 2 feet on center. The flashing is set on top of the second course of the inside wythe and bent down to go under the bed mortar of the outside wythe. Weep holes are installed directly over the flashing so that any accumulated water flows easily (Fig. 7-18 and Fig. 7-19).

Using a cavity wall with a minimum 3-inch cavity would provide far better thermal insulation than does a solid wall using approximately the same amount of brick and mortar. Adding an appropriate type of insulation can increase the thermal resistance phenomenally. There are several kinds of insulating material that can be used in a brick cavity wall, but any insulating should not interfere with the flow of moisture collecting on the interior of the outside wythe. The insulation itself must be unaffected by accumulated moisture, both in insulating properties and durability. Most modern insulations are pretty well impervious to water damage. Any granular fills used

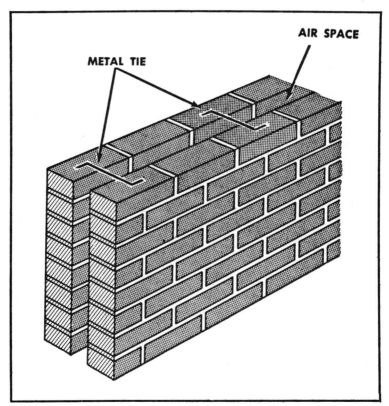

Fig. 7-19. A rowlock brick wall.

should not be the type that will settle, as settling will leave the upper portion of the wall uninsulated. Inorganic materials of adequate fire resistance, as well as vermin and rot resistance, are required. Suitable materials include granular water resistant vermiculite or silicone treated perlite. Rigid board insulation of the plastic foam type is also suitable and, though expensive, offers somewhat higher R values for lesser thickness than do most other types of insulation.

Sound transmission of insulated cavity walls is very low. Traditionally, the best way to cut down on sound transmission has been to build discontinuous units during construction. The second best has been to use massive construction methods. With a brick cavity wall, both traditions are well met. Should someone rap the outer wythe with a hammer, for example, the sound will travel quickly to the inside of the wythe. There it will meet nothing but dead air. Insulation materials themselves do not provide the actual insulation. The insulating value is provided by the lack of movement of the air trapped by the insulation. This fact or works just as well for sound transmission as for heat transmission.

Metal ties must be used to tie together the wythes of cavity walls. These are placed every sixth course, on 2-foot centers (Fig. 7-20). According to the Brick Institute of America, such ties should comply with ASTM A82 or A185, and the ties should also have a corrosion resistant coating such as zinc. Occasionally, stainless steel ties are used, but these do tend to increase the cost of the structure (Fig. 7-21).

The mortar used for cavity walls would be the same as used in all brick wall construction, with some emphasis on going to type S mortar where winds could possibly exceed 80 miles per hour. Flashing is best formed from sheet aluminum, since plastics and other such materials are likely to fail. Having to replace the flashing in this type of construction would be a very expensive job and an awful lot of work. It pays to start with the best and then not have to worry.

Bonds used for cavity walls are the same as for any other type of brick construction, with probable emphasis on common, or American bonds and the runnning bond. With such a cavity wall, the header courses would be half bricks, or bats, so as to leave the cavity as open as possible. The cavity, in case, should never be less than an inch deep.

Detailing is rather important in any brick construction. It is even more important in cavity wall construction since such walls are almost always a part of a home. Bonding is provided in the usual

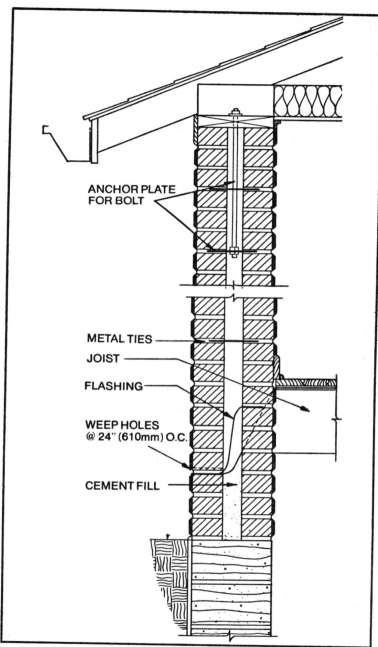

ANCHOR PLATE
FOR BOLT

METAL TIES

JOIST

FLASHING

WEEP HOLES
@ 24" (610mm) O.C.

CEMENT FILL

Fig. 7-20. A cavity wall.

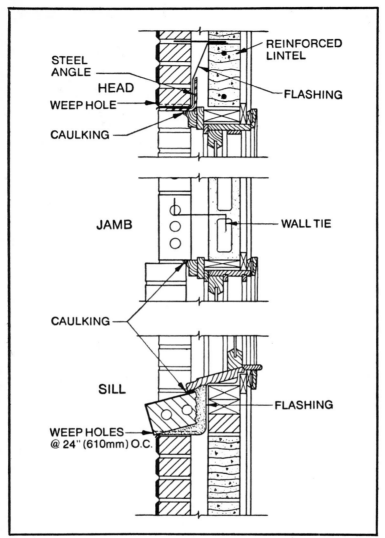

Fig. 7-21. Cutaway of a window head, jamb and sill.

ways, with the addition of the metal ties. The Brick Institute of America recommends three-sixteenth of an inch diameter corrosion resistant ties of sufficient length for each end to be bent 90 degrees, leaving a hook of at least 2 inches. There should be at least one tie for each 4½ suqare feet of wall area, with ties no further apart than 24 inches on the horizontal line and 3 feet vertically. Additional ties are needed within a foot of all openings and spaced no less than every 3 feet around the perimeter of the wall.

Weep holes are essential. They are run above every strip of flashing, no further apart than 2 feet on center. Otherwise, the interior wall will be constantly damp. All caps, copings and lintels must be flashed and have weep holes provided.

When selecting doors and windows for cavity wall construction projects, you'll first need to know exactly how thick your wall is going to be. It is easily possible to set a window back into the wall, but having the window only a standard 5 inches thick set into a foot thick cavity wall looks a little out of place.

Windows with wider frames are available on special order at slight extra cost. The cost of wood frame windows of good quality is quite high but it doesn't pay to go with less than the best you can

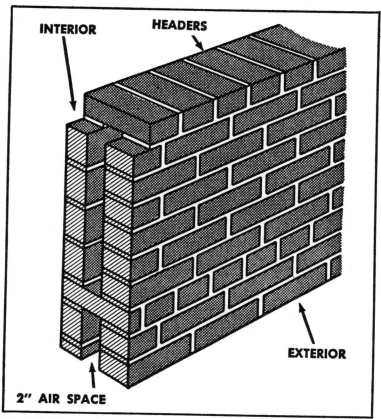

Fig. 7-22. Details of a rowlock backwall.

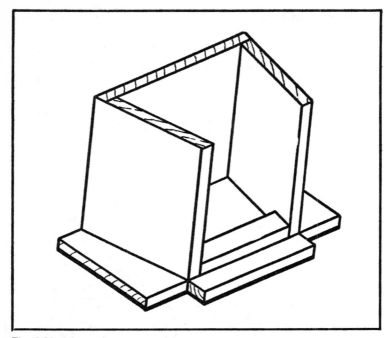

Fig. 7-23. A beam box.

afford since double lazing is pretty much an essential and triple glazing is even more preferable. In many cases, simply adding an aluminum storm window to a double glazed window will provide excellent additional protection against heat loss. Therefore, buying triple glazed units at the outset is not essential when you must save money.

A rowlock wall is another type of cavity wall, although it is not as watertight as the standard cavity wall. Because of the header courses, heat loss is greater. The face of the rowlock wall appears the same as a common bond wall with a header course every seven courses or so. One wythe of bricks is run as rowlock so that the header bricks will be long enough to form a structural bond between the two wythes. A second type of rowlock wall has both wythes run as rowlocks. The first type provides you with about a 2-inch cavity wall, while the second has about a 4-inch cavity (Fig. 7-22).

As with solid walls, brick is laid in the same manner for cavity and rowlock walls. All the procedures of joint making and closures are followed and the corner leads are run first, with the wall erected between corner leads afterwards.

BRICKS AND BEAMS

When beams are to be set in brick walls, you'll need to take a few extra steps. When a brick house is being constructed—not just brick veneer, but all brick—you'll need to set beams and joists into the brick walls in order to have floors and ceilings. In most cases, cavity walls will be used for this sort of construction in order to provide proper insulation values. In order to keep mortar away from the joists, it is best to construct a wall box of pressure treated lumber to hold the end of the joist or beam. The mortar can eventually cause dry rot in the wood if contact is maintained and that portion of the wall becomes damp (Fig. 7-23). Beams are cut at an angle sufficient to allow it to drop free of the wall in case of fire. The reasoning is simple. If the beam is set so that it can drop free of the wall when the interior of the house burns, the wall above the beam won't be damaged by the falling material nor should the wall below be hurt. Therefore, the house, assuming the brick itself is not badly damaged, can be more easily re-built. The beam will bear on the full width of the inside wythe of brick. It is best not to use rowlock wythes for the bearing segment of the wall (Fig. 7-24).

Joist anchors are placed at every fourth joist and three-sixteenth of an inch wall ties should be placed in the first course below the joists, with a metal anchor used for lateral support every 8 feet along the joist. The joists above the metal anchors require solid bridging, as you can see in Fig. 7-25.

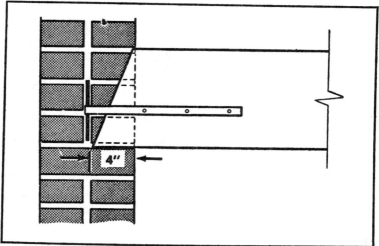

Fig. 7-24. A joist angle cut.

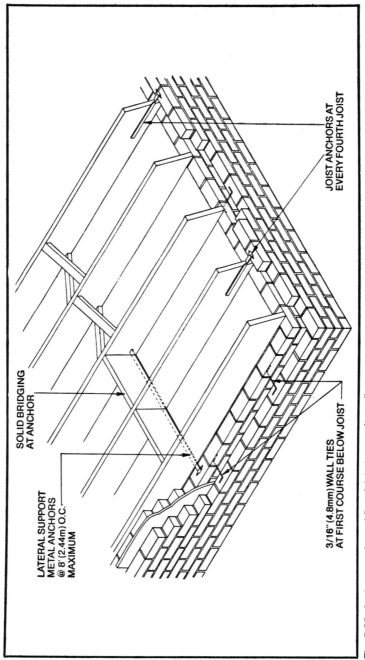

JOIST ANCHORS AT EVERY FOURTH JOIST

SOLID BRIDGING AT ANCHOR

LATERAL SUPPORT METAL ANCHORS @ 8' (2.44m) O.C. MAXIMUM

3/16'' (4.8mm) WALL TIES AT FIRST COURSE BELOW JOIST

Fig. 7-25. Anchorage of wood floor joists to a cavity wall.

Fig. 7-26. One type of single wythe wall design.

149

For simple partition walls, indoors or out, a single wythe brick wall is quite acceptable (Figs. 7-26 and 7-27).

THIN BRICK WALLS

Building interior or exterior walls of a single wythe of brick is an excellent way of providing decoration, windbreaks, area separation and a generally more attractive look. Interior walls should be built on well supported floors only, since even a single row of brick can be quite heavy if the structure is of any size at all. This might mean, if you're building on a wooden first floor, adding a pier or two to your basement support system. In most cases, it would be best to check with a structural engineer first since that cost will be a lot less than having the whole mass drop down on the furnace. For exterior work, the basics of wall building are followed just as with other brick walls. In some instances you'll find it possible to use piers at varied points along the brick wall rather than laying a footing for the entire length of the wall. However, I'm not fond of this method in areas where the frost depth reaches much over 1 foot. Even with good piers, there is still a possibility of some frost heave cracking the wall. The longer the wall, the greater the chances it will eventually crack in the manner.

If the wall is reinforced with steel, then piers become more practical. The Brick Institute of America has supplied recommended reinforcing charts and details for such cases. The pier foundation chart is for pier diameter, not pier depth (Tables 7-7, 7-8 and 7-9). The depth of any foundation work is always determined by

Table 7-7. Reinforcing Steel.

Wall Span, ft	Vertical Spacing, in.								
	Wind Load, 10 psf			Wind Load, 15 psf			Wind Load, 20 psf		
	A	B	C	A	B	C	A	B	C
8	45	30	19	30	20	12	23	15	9.5
10	29	19	12	19	13	8.0	14	10	6.0
12	20	13	8.5	13	9.0	5.5	10	7.0	4.0
14	15	10	6.5	10	6.5	4.0	7.5	5.0	3.0
16	11	7.5	5.0	7.5	5.0	3.0	6.0	4.0	2.5

Note:

A = 2 - No. 2 bars
B = 2 - 3/16-in. diam wires
C = 2 - 9 gage wires

Reinforcing steel to be placed in mortar joints may be any one of three types, as identified under A, B, C. They are manufactured by many companies under trade names. Determine cost and availability in your area, then select one.

Fig. 7-27. A geometrical wall pattern.

Table 7-8. Pier Reinforcing Steel.

Wall Span, ft	Wind Load, 10 psf			Wind Load, 15 psf			Wind Load, 20 psf		
	Wall Height, ft			Wall Height, ft			Wall Height, ft		
	4	6	8	4	6	8	4	6	8
8	2#3	2#4	2#5	2#3	2#5	2#6	2#4	2#5	2#5
10	2#3	2#4	2#5	2#4	2#5	2#7	2#4	2#6	2#6
12	2#3	2#5	2#6	2#4	2#6	2#6	2#4	2#6	2#7
14	2#3	2#5	2#6	2#4	2#6	2#6	2#5	2#5	2#7
16	2#4	2#5	2#7	2#4	2#6	2#7	2#5	2#6	2#7

(1) Within heavy lines 12 by 16-in. pier required. All other values obtained with 12 by 12-in. pier

your locality's frost depth. Any building inspector should be able to provide you with this information.

The basics are simple. Stake out the fence perimeter and the pier locations. Position the vertical steel rod, of the size the chart indicates, in the pier holes and fill the holes with concrete. Maintain as close to a plumb as possible with the vertical steel. Place the bed mortar directly on the ground and lay the first course of bricks in this. To make laying the first course easier, scrape or tamp to get the ground as level as possible or cut away to make a sort of step action. The first brick course then has the reinforcing steel laid directly on it and the mortar is placed for the next course. Mortar should be just a bit less thick than normal so that it will flow well around the horizontal steel and give a good bond.

To save yourself a lot of trouble, make a plan before you start. Take careful measurements and decide just where your piers are going to go before digging holes and pouring concrete. Changes

Table 7-9. Embedment Chart for Pier Foundation.

Wall Span, ft	Wind Load, 10 psf			Wind Load, 15 psf			Wind Load, 20 psf		
	Wall Height, ft			Wall Height, ft			Wall Height, ft		
	4	6	8	4	6	8	4	6	8
8	2'-0"	2'-3"	2'-9"	2'-3"	2'-6"	3'-0"	2'-3"	2'-9"	3'-0"
10	2'-0"	2'-6"	2'-9"	2'-3"	2'-9"	3'-3"	2'-6"	3'-0"	3'-3"
12	2'-3"	2'-6"	3'-0"	2'-3"	3'-0"	3'-3"	2'-6"	3'-3"	3'-6"
14	2'-3"	2'-9"	3'-0"	2'-6"	3'-0"	3'-3"	2'-9"	3'-3"	3'-9"
16	2'-3"	2'-9"	3'-0"	2'-6"	3'-3"	3'-6"	2'-9"	3'-3"	4'-0"

(1) Within heavy lines 24-in. diam foundation required. All other values obtained with 18-in. diam foundation

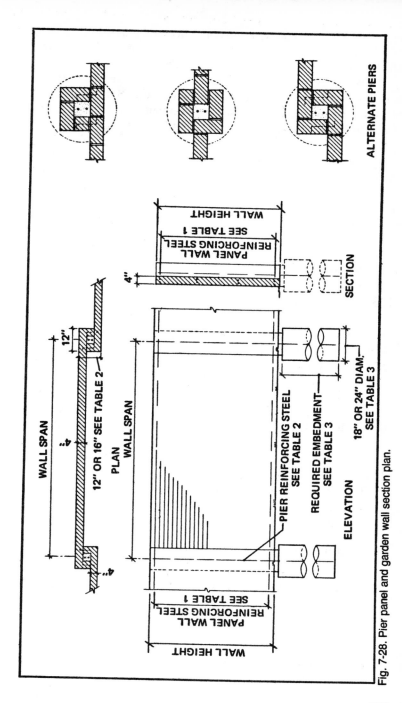

ALTERNATE PIERS

PLAN

WALL SPAN

12" OR 16" SEE TABLE 2

WALL SPAN

4"

12"

SECTION

PANEL WALL
REINFORCING STEEL
SEE TABLE 1

WALL HEIGHT

4"

WALL HEIGHT

PANEL WALL
REINFORCING STEEL
SEE TABLE 1

PIER REINFORCING STEEL
SEE TABLE 2

REQUIRED EMBEDMENT
SEE TABLE 3

18" OR 24" DIAM.
SEE TABLE 3

ELEVATION

Fig. 7-28. Pier panel and garden wall section plan.

153

made on paper, as you walk the perimeter of the fence, are much more cheaply made than are changes in design after construction starts. They're also a lot easier (Fig. 7-28).

Without using reinforcing steel, you can lay a single row brick wall by using a good footing and foundation. Use 8-inch concrete block for any foundation walls, to save money, and use poured concrete for a one-sixteenth inch by eight-inch thick footing. If the wall will be very tall and the load bearing capacity of the soil in your area is poor, float the wall on a 2-foot wide footing some 12 inches thick. With the reinforced wall, you can use any bond pattern you prefer including stacked bond. With a straight wall, you must use one of the bonding styles that includes more structural integrity. Probably the most popular are the running bond, and its variations, and the common bond (Fig. 7-29).

In both cases, use type S mortar to increase resistance to wind loading. It is an unpleasant thing to spend weeks getting a wall up only to have a gale come up and lay it in a heap.

SERPENTINE WALLS

A clever design technique that provides lateral strength to a relatively thin wall is employed for building serpentine walls. Serpentine walls have a long tradition of usage. Well known examples can be found in Williamsburg, Virginia and at the University of Virginia in Charlottesville (Fig. 7-30).

The proper curvature of a serpentine wall is extremely important since the strength of the wall is primarily derived from the curve. Always keep in mind that the radius of the curvature of a 4-inch wall should not be more than twice the height of the finished wall. The depth of the curvature should not be less than one-half the height of the finished wall (Fig. 17-31).

This rule is especially important for patio or garden walls that are 4 inches thick. For loadbearing walls such as high retaining wall or high garden walls (Fig. 7-32), an engineering design is required.

Structural Design

Use the following formula to determine the geometrical properties of a non-loadbearing serpentine wall that is 4 inches thick:

$$f_m = \frac{P}{A} \pm \frac{Mc}{I}$$

where:
f_m = extreme fiber stress (psi)
M = bending moment (inches-pounds)

Fig. 7-29. One of many popular brick wall designs.

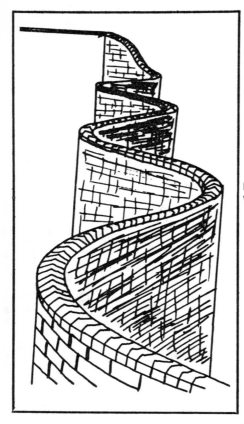

Fig. 7-30. A brick serpentine wall.

P = compressive load (pounds)
A = area (square inches)
c = distance from the centroidal axis to the extreme fiber (inches)
I = moment of inertia about the centroidal axis (inches⁴)

Working Stress

The design of a non-loadbearing wall is controlled by its resistance to flexural tensile stress caused by lateral loads. See Table 7-10 for the allowable tension in flexure and shear stresses for non-reinforced masonry. If the stresses shown in Table 7-10 are exceeded, the masonry can be reinforced with steel dowels. If the base of the garden wall is constructed on a concrete foundation, the allowable stresses found in Table 7-10 are not applicable. However, be sure to keep in mind the design requirement that the overturning moment not exceed two-thirds of the righting moment.

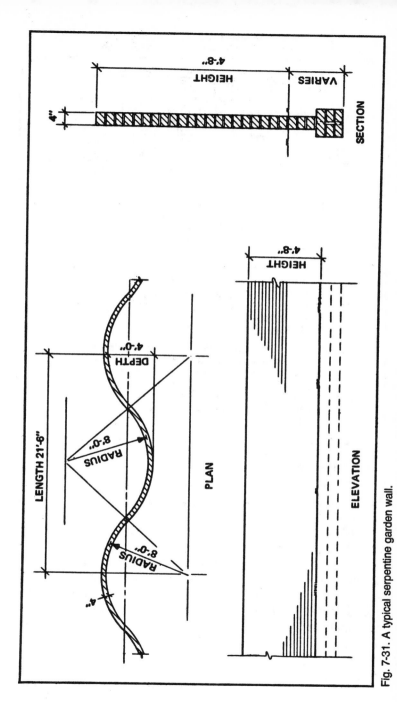

Fig. 7-31. A typical serpentine garden wall.

157

Table 7-10. Allowable Stresses in Tension in Flexure and Shear in Non-reinforced Brick Masonry[1]

Construction	Allowable Stresses, psi					
	Tension in Flexure				Shear	
	Normal to Bed Joints[2]		Parallel to Bed Joints[3]			
	Mortar Type [4]				Mortar Type [4]	
	M or S	N	M or S	N	M or S	N
With inspection	36	28	72	56	50	40
Without inspection	24	19	48	37	33	26

(1) When there is no engineering or architectural inspection to insure that the workmanship requirements of the SCPI Standard are satisfied, the values given for brick masonry without inspection shall be used.

(2) Direction of stress is normal to bed joints; vertically in normal masonry construction.

(3) Direction of stress is parallel to bed joints; horizontally in normal masonry construction. If masonry is laid in stack bond, tensile stresses in the horizontal direction shall not be permitted in the masonry.

(4) Mortar for use in non-reinforced brick masonry shall conform to Standard Specifications for Mortar for Unit Masonry ASTM C 270-64T, type M, S or N, except that it shall consist of a mixture of portland cement (type I, II or III), hydrated lime (type S) and aggregate, where values given in this table are used.

Geometrical Properties

The geometrical properties of a serpentine wall (Fig. 7-33) can be defined by assigning r as the radius of the curvature to the center line of the wall; Φ as the angle between the radial line perpendicular to the centroidal axis and the line connecting the two centers of curvature; and t as the thickness of the wall.

Equations for determining Φ and r in terms of t and in terms of L, the length of the repeating section along the centroidal axis, and in terms of c, the perpendicular distance from the centroidal axis to the extreme fiber of the wall, are given below. In these equations, r, t, c and L are expressed in inches and Φ is expressed in radians:

$A = 4 \, \Phi \, r \, t$

$L = 4 \, r \, \text{Sin} \, \Phi$

$L = 4 \, r \, \Phi$

Fig. 7-32. A loadbearing serpentine wall.

$$c = r + t/2 - r \ \text{Cos} \ \Phi$$

$$r = \frac{I}{2} \left[\frac{L^2}{16 \ (c-t/2)} + (c - t/2) \right]$$

$$\Phi = 2 \left[\text{Cot}^{-1} \frac{L}{4(c - t/2)} \right]$$

$$I_{CA} = \frac{A}{2} \left(r^2 X + \frac{t^2}{4} Y \right)$$

$$S_{CA} = \frac{A}{2c} \left(r^2 X + \frac{t^2}{4} Y \right)$$

$$K_{CA} = \left[\frac{1}{2} \left(r^2 X + \frac{t^2}{4} Y \right) \right]^{\frac{1}{2}}$$

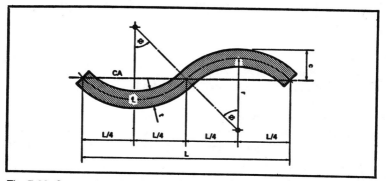

Fig. 7-33. Geometrical properties of a repeating section of a serpentine wall.

159

where:

A = cross-sectional area of the repeating section (square inches).

L_t = length along the center line of the repeating secion (inches)

I_{CA} = moment of inertia about the centroidal axis (inches^4)

K_{CA} = radius of gyration about the centroidal axis (inches)

S_{CA} = section modulus about the centroidal axis (inches^3)

$$X = 1 - \frac{3 \, Sin \, \Phi \, Cos \, \Phi}{\Phi} + 2 \, Cos^2 \, \Phi; \text{ (see Table 7-11)}$$

$$Y = 1 - \frac{Sin \, \Phi \, Cos \, \Phi}{3} \text{ (see Table 7-11)}$$

Bending Moment

Several types of force, such as wind, an earthquake or an explosion, can produce lateral loads on a non-bearing wall. The result is know as the bending moment. The bending moment and the point of maximum moment depend on the design of the wall. Design loads and wind loads vary with the height of the structure, geography and local building codes.

If the wall is supported only at its top and bottom, the maximum moment occurs at the middle of the wall. To compute the maximum moment, use the following equation:

$$M = \tfrac{1}{8} \, wh^2 \, L$$

where:

M = moment (inches-pounds)

w = lateral load (pounds per square foot)

h = height of the wall (feet)

L = length of the wall along the centroidal axis (inches)

If the wall is supported at its top and merely fixed at the bottom, the maximum moment can be computed with the following equation:

$$M = \tfrac{1}{8} \, wh^2 \, L$$

where:

M = moment (inches-pounds)

w = lateral load (pounds per square foot)

h = height of the wall (feet)

L = length of the wall along the centroidal axis (inches)

If the wall is not supported at the top and it is fixed at the bottom, the maximum moment will occur at the bottom. To compute the maximum moment, use the following equation:

$$M = \tfrac{1}{2} \, wh^2 \, L$$

Table 7-11. Geometrical Coefficients for Serpentine Walls.

Φ Degrees	Φ Radians	Sin Φ	Cos Φ	Cot Φ	X	Y
5	0.08727	0.08716	0.99619	11.43005	0.00002	0.66836
10	0.17453	0.17365	0.98481	5.67128	0.00025	0.67339
15	0.26180	0.25882	0.96593	3.73205	0.00124	0.68169
20	0.34907	0.34202	0.93969	2.74748	0.00387	0.69309
25	0.43633	0.42262	0.90631	2.14451	0.00932	0.70739
30	0.52360	0.50000	0.86603	1.73205	0.01902	0.72434
35	0.61087	0.57358	0.81915	1.42815	0.03457	0.74362
40	0.69813	0.64279	0.76604	1.19175	0.05770	0.76489
45	0.78540	0.70711	0.70711	1.00000	0.09014	0.78779
50	0.87266	0.76604	0.64279		0.13359	0.81192
55	0.95993	0.81915	0.57358		0.18960	0.83685
60	1.04720	0.86603	0.50000		0.25951	0.86217
65	1.13446	0.90631	0.42262		0.34434	0.88746
70	1.22173	0.93969	0.34202		0.44476	0.91231
75	1.30900	0.96593	0.25882		0.56102	0.93634
80	1.39626	0.98481	0.17365		0.69288	0.95917
85	1.48353	0.99619	0.08716		0.83962	0.98049
90	1.57080	1.00000	0.00000		1.00000	1.00000

where:

M = moment (inches-pounds)

w = lateral load (pounds per square foot)

h = height of the wall (feet)

L = length of the wall along the centroidal axis (inches)

Compressive Load

The weight of the wall above a specific section is known as the compressive load for a non-loadbearing wall. Since the design of a non-loadbearing wall is controlled by its flexural resistance, the location of the maximum moment is strategic. The magnitude of the compressive load for the first example shown above can be computed with the following equation:

$$P = \frac{h\,L_t\,t'\,W}{288}$$

where:

P = compressive load (pounds)

h = height of the wall (feet)

L_t = length of the wall along its t (inches)

161

t' = nominal wall thickness (inches)

W = weight of the wall (pounds per cubic foot), assumed to be 120 pounds per cubic foot for brick masonry

In order to find the magnitude of the compressive load which coincides with the maximum bending moment for the second and third examples given above, use the following equation:

$$P = \frac{h\,L_t\,t'\,W}{144}$$

where:

P = compressive load (pounds)

h = height of the wall (feet)

L_t = length of the wall along its L_t (inches)

t' = nominal wall thickness (inches)

W = weight of the wall (pounds per cubic foot), assumed to be 120 pounds per cubic foot for brick masonry

Example

In order to determine the extreme fiber stresses resulting from a 20-psf wind load on a 4-inch serpentine wall, use the formula provided below. For this example, the dimensions of the wall are 8 feet in height, 2 feet 6 inches in total depth and a repeating section every 12 feet.

w = 20 psf

W = 120 lb per cu ft

h = 8 ft 0 in.

t = 4 in. nominal = 3⅝ in. actual

L = 12 ft 0 in. = 144 in.

$c = \dfrac{2 \text{ ft } 6 \text{ in.}}{2} = 15$ in.

$$\Phi = 2\left[\, \mathrm{Cot}^{-1} \frac{144}{4\left(15 - \dfrac{3.625}{2}\right)} \right]$$

$$= 2\,[\ \mathrm{Cot}^{-1}\ 2.74\ ]$$

$$= 2\,[\ 20 \deg\,]$$

$$= 40 \deg = 0.69813 \text{ radians}$$

$$r = \frac{144^2}{32\left(15 - \frac{3.625}{2}\right)} + \left(\frac{15 - \frac{3.625}{2}}{2}\right)$$

$$= 56 \text{ in.}$$

$$A = 4 \ (0.69813) \ (56) \ (3.625)$$

$$= 567 \text{ sq. in.}$$

$$I_{CA} = \frac{567}{2} \left[\ (56)^2 \ (0.05770) \ + \ \frac{(3.625)^2}{4} \ (0.76489) \right]$$

$$= 52,100 \text{ in.}^4$$

$$L_{\mathfrak{k}} = 4 \ (56) \ (0.69813) \ = \ 156.5 \text{ in.}$$

$$P = \frac{8 \ (156.5) \ (4) \ (120)}{144} \ = 4175 \text{ lb}$$

$$M = \tfrac{1}{2} \ (20) \ (8)^2 \ (144) \ = \ 92,200 \text{ in-lb}$$

$$f_m = \frac{4175}{567} \ \pm \ \frac{92,200 \ (15)}{52,100}$$

$$f_m = 7.4 + 26.6 = 34 \text{ psi (max)}$$

$$f_m = 7.4 - 26.6 = -19.2 \text{ psi (min)}$$

Cleaning & Repair

Properly constructed brick walls, barbecues, patios and other structures require very little care at all. An old brick wall that is in rough shape can sometimes be replaced less expensively than it could be repaired. Generally, the cost without labor isn't all that great, but the work is quite tedious. It involves, most often, chiseling out old mortar and going through the tuckpointing procedures to get a new bond and mortar joint. The tedium of the repair job is one reason I've laid so much emphasis on doing it properly, with the best materials possible. Brick walls have stood for thousands of years in some environments and are still in fairly good shape. Of course, with industrial and other types of pollution abounding today, there is liable to be more wear and tear than in centuries gone by. I understand that some artifacts , such as Cleopatra's Needle, have undergone more surface destruction in the past 20 years than in almost a dozen centuries before. This is simply because of the contaminants in the air.

Tuckpointing is described in the previous chapter. It is a job that requires a fair amount of effort if done over large areas and it is a good idea to use the pointing trowel to lay in mortar in one-quarter inch layers instead of all at once.

Repairing cracked brick is another matter. It is done in just about the same manner as you would repair concrete. Chip out—gently to avoid enlarging the crack—at right angles to the surface or with a slight undercut a good line and clean it thoroughly. The cut into the crack should be at least half an inch deep to allow for enough mortar to get a good bond. The edges must not be feathered as these would then chip away very rapidly. Dampen the brick and allow it to almost surface dry. Use a general purpose mortar. You can add a stain to approximate the color of your brick if you prefer. Water base stains are available in many colors. Let the mortar come close to taking its initial set before applying it to the cracked brick. This procedure will help reduce shrinkage. If the mortar is correctly mixed, waiting 45 minutes to an hour will be more than sufficient to reduce shrinkage. Smooth the surface to as close to the texture of the brick as you can.

NEW BRICK

While a skilled professional brick mason can lay a brick wall with few or no mortar stains on the brick, the rest of us are seldom so lucky as to get away with only a few dabs here and there. Once the job is complete, use an old putty knife to chip away any large chunks of mortar that have adhered to the surfaces. Once this is done, soak the surface of the brick thoroughly. Mix a preparation of 1 part commercial muriatic acid to 9 parts of water. Make sure to pour the acid *into* the water. Put on goggles, rubber gloves and wear a long sleeved shirt while using this solution. Use a long handled, stiff fiber brush to cover some 15 to 20 square feet at a time with the muriatic acid. Don't cover any more since the solution will dry before you can rinse it off.

The wall must be well soaked before the acid solution is applied. Otherwise the solution will be absorbed into the brick instead of working on the surface stains. As soon as you've finished the first section of 15 or so square feet, rinse it thoroughly with clear water. There is a need for an immediate and thorough rinsing since the acid will, if left on too long, attack the mortar joints and could possibly ruin the bond.

All wood parts such as door and window frames must be well protected from contact with the muriatic acid. Generally, taping on some polythelene sheeting will do this job well.

Efflorescence will appear on many new brick walls and must be removed. Efflorescence is a whitish deposit and messy looking. It consists of soluble salts contained in the brick that come to the surface when the brick is wet. Once the water evaporates, the mess remains. If the deposits are not extreme, efflorescence can be removed with clear water and a stiff brushing. If that doesn't work, it is time to break out the muriatic acid again.

OLD BRICK

There are several methods used to clean old brick. Some will almost force you to tuckpoint the mortar joints afterwards. Much depends on the condition of the surface at the outset, the type of brick used—smooth brick is a lot more easily cleaned of most grime than is rough brick—and the type of contamination. For brick that has been painted, the need for extreme methods is obvious. In such cases, sandblasting is about the only sane way to do the job. Any extensive surface would be far too much to handle with a wire brush. Although sandblasting roughens the brick surface and will probably present a need for extensive pointing, it does a much better job than wire brushing.

SANDBLASTING

Sandblasting requires special, quite expensive equipment. Fortunately, the equipment can be rented almost everywhere. Compressed air forces the sand through a nozzle and the resulting abrasive stream removes the surface to whatever depth is needed to clean the grit or paint. As I said, it does roughen the surface of the brick and it will change the looks of a smooth brick wall, but if the smooth brick wall has 10 or 15 coats of old paint, it likely looks pretty bedraggled anyway.

After the sandblasting and tuckpointing is finished, you'll find yourself with a surface rough enough to quickly collect a new padding of grime. This is probably the reason the wall was painted in the first place. There isn't much you can do about the surface texture, but a good, transparent brick sealer can cut down on penetration of the grime and make future cleanings an easier job.

STEAMCLEANING

Sandblasting should only be done to a depth that removes otherwise unremovable stains. If further cleaning is needed, steam or water can be useful. The tool for producing the steam/water spray, often mixed with detergent, could set the average person's wallet to whimpering. But it is another that is readily available for rent almost everywhere.

If sandblasting is completed, or not needed, the pressurized steam will lift out and wash away almost all contaminants. Select a steam sprayer with a high velocity and don't worry too much about the volume of water used. The velocity with which the steam impacts the wall, and the dirt in and on it, is of greater importance. A high flow rate can help speed up the job. Look for a machine that will produce from 140 to 150 pounds of pressure per square inch.

Clean only about a 3-foot square section of wall at one time. Then rinse completely with clear water from a garden hose. The use of some cleaning compounds can help lift out embedded grease and other grime. Sodium carbonate and triosodium phosphate are especially recommended for use on brick. If hardened deposits are left after steam cleaning, scrape them off with a wire brush or an old putty knife. Use an old putty knife since as a new one will definitely become old if first used in this manner. Rinse the surface heavily and hit it once more with the steam cleaner.

HYDROCHLORIC ACID AND WATER

Removing some other types of coatings doesn't require the use of a sandblaster. Calcimine, whitewash coatings and some water base paints will come loose if you wash them down with a solution of 1 part hydrochloric acid to 5 parts of water. Use a fiber scrub brush with a very long handle. The solution will foam as it hits the coating and it must be scrubbing away. This solution can be hazardous. Wear goggles, rubber gloves and a longsleeved shirt and then do your best to not splash any on yourself. A complete rinse is essential to remove the acid and the coating scum.

Commercial paint removers are sometimes used to remove paint from smooth brick walls where the smoothness is to be retained. I don't much care to do this, as the work load is immense. You can never quite tell what the chemical composition of the paint remover is. Try it on a small section first, before moving on to the main part of the job. In most cases, sandblasting or some other type of remover is probably better, and cheaper and less work.

Two pounds of trisodium phosphate in one gallon of very hot (not boiling) water makes a reasonable paint remover. Burning paint off with a blowtorch or a propane torch is more practical on brick walls than on frame walls, but I don't like a procedure. It requires a lot of scraping and it is dangerous.

Occasionally, you might find a few other types of stains used on brick masonry. Iron stains, especially in areas where water might run over the surface and contain a great many minerals, can be removed with a solution of 7 parts of pure glycerine, 1 part sodium citrate and 6 parts of lukewarm water. Add whiting to make a thick paste and apply the mixture to the stain with a trowel. Scrape it off when it is dry and wash the surface well with water. You might have to repeat the application several times to remove the stain.

Smoke stains around fireplaces and chimneys can usually be removed with a smooth, stiff paste made of powdered talc and trichloroethylene (equal parts). Apply, let dry, remove and rinse. Keep the container covered or the trichloroethylene will evaporate. Also, make sure that there is plenty of ventilation as the fumes are far from being healthy. After several tries, if this doesn't take out the stain, dissolve 2 pounds of trisodium phosphate in 5 quarts of water. Then use an enameled pan (and make sure the enamel is not chipped) to mix a dozen ounces of chloride of lime in just enough water to form a thick paste. Use a 2 gallon stoneware jug to mix the two solutions together. Let the lime settle out and draw off the clear liquid and cut it with an equal amount of water. Use powdered talc to make a thick paste and trowel the mixture on. Let dry, scrape off and rinse well.

Oil stains could prove a problem on some brickwork. Again, trisodium phosphate works. One pound of trisodium phosphate is added to a gallon of water and then thickened to a paste with whiting. It is then troweled over the oil stain in a layer at least half an inch thick. After 24 hours of drying time, the paste is removed and the surface rinsed well with clear water. The stain should be gone.

9

Brick Veneer

Most brick construction for new homes today involves brick veneering. Veneering is the building of a wall of another material as the basis for the home and the attachment of a single wythe of brick to face that wall. In most cases, standard wood framing techniques are used to form the back-up wall. Occasionally, metal studs are used and sometimes even concrete block is used. However, using concrete block tends to cut the advantages of having a studded wall behind the brick as an insulation holder. The single wythe of brick is held to the brick (Fig. 9-1).

Brick veneer walls offer several advantages over conventional exterior finishes, solid brick or brick cavity walls. First, the veneer will be durable as only brick can be and it is fire resistant. Brick veneer walls are drainage style walls, so they offer about the maximum possible resistance to water penetration under almost any conditions. The required 1-inch air space between the brick veneer wall and the back-up wall offers a slight R value insulation increase. Also, sound transmission is cut because of the discontinuous construction required. Cost for brick veneer is higher than it would be for standard exterior clapboard or panelling, but the need for maintenance over the years is cut to almost nothing. Brick veneer walls go up more rapidly than do solid or cavity brick walls and material is saved since only a single wythe is necessary. Therefore, the cost is a great deal lower than either other type of brick construction. Brick veneer doesn't need to be restricted to new home construction. It is possible to veneer much of your home, probably for a bit less than the cost of a top quality aluminum or vinyl siding job. More on existing home brick veneering later in the chapter.

For new home construction, the house will have been designed with such a cladding in mind. Your first job is to select the correct brick. The Brick Institute of America recommends a severe weather grade (SW) facing brick because the brick is to be isolated from the back-up wall and will be subjected to all kinds of temperature and other weather extremes. BIA recommends a type N mortar, since it is resistant to weathering under constant exposure to the worst possible conditions.

FOUNDATION

Foundation construction for a brick veneer wall will, in a large part, depend on your local building codes. In no case should the foundation wall be less than the thickness of the brick to be supported, plus the thickness of the interior framing. A few building codes do permit the use of an 8-inch foundation wall (under single family dwellings) if the top of the wall is corbeled no more than 2 inches (Fig. 9-2). To me, it seems wiser to go with a wider concrete block or wider poured wall. Concrete block is readily available in 12-inch nominal widths. This will readily support a 4-inch brick wall, allow room for the minimum of 1-inch between the brick and its back-up wall and allow a sheathed back-up wall (with interior finish materials in place) 5 inches thick. Actually, you've got a couple of inches left over and you might even opt for a 2 × 6 frame. This allows you to open out to 2 feet on center, so that materials costs are not really all that much greater than they are with 2 × 4 studs on 16-inch centers. The heavier house framing members will allow you to just about double the thickness of the insulation you use in the home. The smaller number of studs used also helps cut heat loss. Many people don't realize that a fair amount of heat is lost through the studs used in framing a house.

If you opt for the 8-inch method, the total corbel cannot exceed 2 inches. The individual corbels can't project more than a third of the height of the unit, while the top corbel course must be no higher than the bottom of the floor joist. This final corbel course should also be a full header course.

HEIGHT LIMITS

Even with proper foundations, there are height limits for single wythe brick veneer walls. For most construction in residences,

Fig. 9-1. Brick walls leading to a brick veneer home.

Table 9-1. Height Limitations.

Nominal Thickness of Brick Veneer, (in.)	Stories	Height at Plate, (ft.)	Height at Gable, (ft.)
3	2	20	28
4	3	30	38

these limits are of little hinderance since even when using 3-inch brick you can go up 2 stories, or 20 feet. A full 4-inch brick allows you to raise a wall 30-feet high (Table 9-1).

Once the brick and the materials for the motor are delivered, you'll need to choose a bond style for your wall. In general, any bond can be used, but to aid structural integrity I would avoid using a stack bond unless reinforcing steel is used. However, this is not considered essential by most builders. I just get nervous using a bond that has no inherent structural use. This has something to do with the old feeling of "if it ain't practical, it ain't pretty." As I indicated earlier, I find the stack bond boring anyway.

The ties used to connect the brick wythe to the back-up wall are important to the final strength of the job. For connections to wood frame back-up walls, corrugated metal ties are needed. These should be at least 22 gauge metal, galvanized for corrosion resistance, at least seven-eighths of an inch wide and 6 inches long. Eight-penny nails are used to attach them to the back-up wall. To obtain the greatest holding power, it is best if each tie is nailed so that the nail penetrates a stud or a cat in the wall. It also helps to use ring shanked nails. If you use an extra thick sheathing on the back-up wall, you might have to go to a longer nail since 1 ½ inches of penetration is needed. For metal stud back-up walls, corrosion resistant wire ties are used. These should have at least a 9-gauge diameter and they are attached to the metal studs with corrosion resistant self tapping metal screws. Such screws are available with special heads that allow you to chuck a driver in an electric drill and save a lot of time and energy (Figs. 9-3 and 9-4).

If your back-up wall is concrete block, I would recommend the use of DESA's power hammer. Check local codes since this machine uses .22 caliber blanks to fire the nail into the concrete. Sold under DESA's Remington brand name, this tool is easy to use. It is much quicker and simpler than driving masonry nails. The nail is loaded in the muzzle, the cartridge (in one of four power loads), is

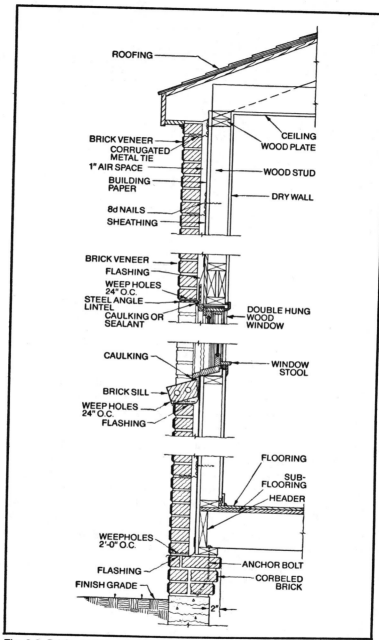

Fig. 9-2. Brick veneer wall, with wood backup wall.

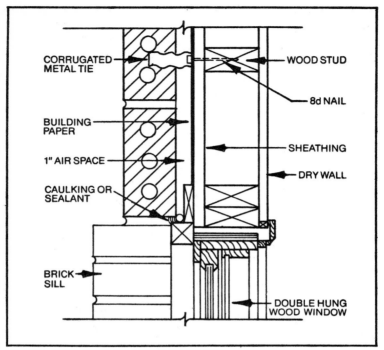

CORRUGATED METAL TIE

BUILDING PAPER

1" AIR SPACE

CAULKING OR SEALANT

BRICK SILL

WOOD STUD

8d NAIL

SHEATHING

DRY WALL

DOUBLE HUNG WOOD WINDOW

Fig. 9-3. Jamb details for a brick veneer wall.

inserted into the chamber and the tool is placed against the surface. This closes the chamber and compresses the safety spring near the muzzle. You then give the handle end of the power hammer a sharp rap with a 16-ounce hammer and the nail is in straight and unbent. That is a lot more than you can say for driving masonry nails.

Not too long ago, I started framing a basement and I used the power hammer to nail the sole plates to the concrete floors. Before the studs were in, we decided to change the floor layout. It took a good bit of yanking with a large crowbar to lift the sole plates. We were thankful to note that the holes into the concrete had very little chipping around the edges, chipping that is characteristic when masonry nails are hammer driven. I then used the power hammer to place ties (of the style used for wood) on a concrete clock chimney to be faced in brick. Again, it worked well.

My only objection to Remington's power hammer is it's cost. Not the approximately $30 the tool itself costs, it is well worth that but the cost of the cartridges. I can understand the nails costing a fair amount since they are specially hardened steel. But I can't understand the cartridges being near $10 for a hundred. Since the ammunition for my .22-caliber pistol costs only about $2.50 for a hundred, the smaller and less complex power hammer cartridges seem overly expensive.

Veneer ties are spaced at intervals of 16 inches horizontally and 2 feet vertically, with no less than 2 inches of the ties imbedded in the mortar joint.

Standard brick wall construction techniques are used for veneer walls, with special care being taken to keep the 1 inch or more gap between the veneer wall and the back-up wall clear of debris of any kind. At the same time, full mortar joints are essential to a good job. Clean up the back of each brick, as it is laid is a good idea to keep mortar out of the air gap.

FLASHING

Flashing and weepholes are used to drain off any water which might penetrate the wall. Flashing is best made of sheet metal. My preference goes to aluminum because it is easy to work with,

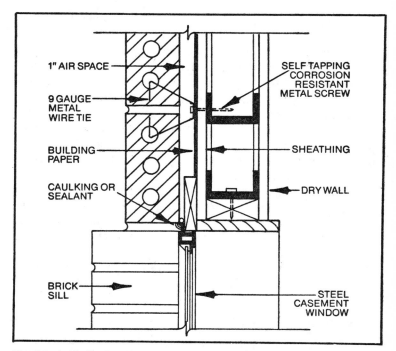

Fig. 9-4. Jamb details with a metal stud back up. Metal screws are used to hold ties.

corrosion resistant and readily available. Most codes will also accept such things as sheet plastics and bituminous membranes for flashing. These can save a few dollars for you but the amount of money spent on flashing is such a small part of the entire materials cost for any brick work, it doesn't seem especially sensible to scrimp at all on such an important point.

Flashing is secured to the back-up wall and then run to the top of an above grade course of brick. Any brickwork running below the flashing should be fully grouted to the height of the flashing. Weepholes are positioned directly above the flashing and spaced on 2-foot centers. Weepholes can be formed in several ways. Part or all of a head joint can be left out or a forming material can be used, such as an oiled rod. In some cases, plastic tubes or rope wicks are used and are left in place after the wall is completed. Of the two, I think the plastic tube is the best bet (Fig. 9-5).

WINDOWS AND DOORS

For the insertion of windows and doors in new brick veneer construction, the Brick Institute of America recommends the use of steel lintels that are at least one-quarter inch thick. Each horizontal leg should be least 3 ½ inches wide—for nominal 4-inch brick—or 2 ½ inches wide—for nominal 3-inch brick.

Sealing around doors and windows is of great importance in preventing both air and water leaks. Outside joints around window and door frames require good accuracy, as they must be at least one-quarter of an inch wide and shouldn't exceed three-eighths of an inch in width. These are cleaned to a depth of three-quarters of an inch and then filled with a top quality silicone sealant forced in with a pressure gun to get a tight and long lasting bead. If the joint is deeper than three-quarters of an inch, use a compressible backing rope as fill before caulking.

In no case does it pay to scrimp on these materials. Good caulking will last many years and earn its initial cost in fewer problems and less work. Also, the possibility of unnoticed leakage causing damage to a door or window frame is cut down when good quality, flexible and durable sealants are used for caulking.

Mortar joints for brick veneer construction should be either V-shaped or concave and it should be as well made and tightly made as possible.

EXISTING CONSTRUCTION

It is possible to gain the advantages of brick veneer construction in houses already standing, without much more difficulty than

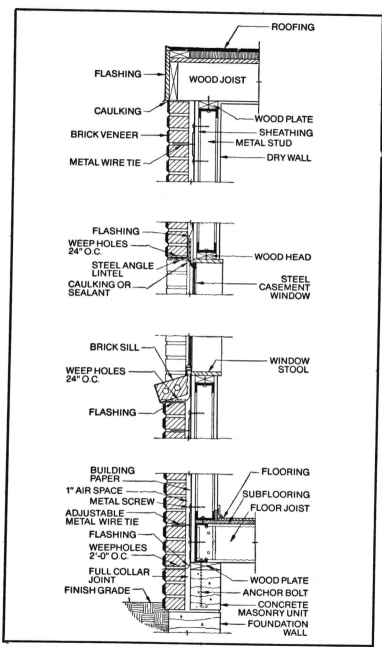

ROOFING

FLASHING

WOOD JOIST

CAULKING

WOOD PLATE

BRICK VENEER

SHEATHING

METAL STUD

METAL WIRE TIE

DRY WALL

FLASHING

WEEP HOLES 24" O.C.

WOOD HEAD

STEEL ANGLE LINTEL

CAULKING OR SEALANT

STEEL CASEMENT WINDOW

BRICK SILL

WINDOW STOOL

WEEP HOLES 24" O.C.

FLASHING

BUILDING PAPER

FLOORING

1" AIR SPACE

SUBFLOORING

METAL SCREW

FLOOR JOIST

ADJUSTABLE METAL WIRE TIE

FLASHING

WEEPHOLES 2'-0" O.C.

FULL COLLAR JOINT

WOOD PLATE

FINISH GRADE

ANCHOR BOLT

CONCRETE MASONRY UNIT

FOUNDATION WALL

Fig. 9-5. Details of a brick veneer wall using a metal stud back up.

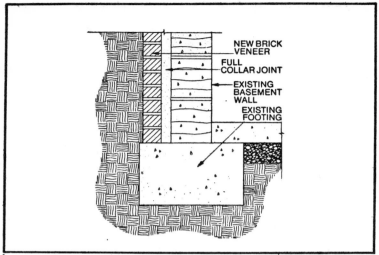

Fig. 9-6. Footing used for brick veneering construction.

you would have in placing the veneer wythe on a new home designed to accept it. While the costs might include some fairly expensive excavation in some cases, in others the existing foundation will serve very well. In still other cases, special steel plate supports are available where digging would be too expensive or difficult. Much depends on the existing foundation and the ground conditions.

Brick veneer should not be used as a support for walls that are not in good structural condition before the installation of the veneer. It doesn't matter at all how the walls that are to be covered look, but you do need a solid back-up wall of good strength. A single wythe wall of nominal 3- or 4-inch brick that is not reinforced simply does not have the structural strength to hold up an already collapsing house. The purpose of adding brick veneer to an already existing house is primarily to cut down on exterior maintenance. There are added benefits of beauty, the greater insulation provided by the air gap, the reduction of sound transmission and, usually, an enhanced home value.

Height limitations for adding brick veneer to older homes of frame construction are the same as for brick veneer in new construction. Brick and mortar selections are the same, as are bond styles.

Foundation requirements differ. If your existing foundation footing, after excavation, is as wide or wider than the brick, plus the minimum 1-inch air gap, then the veneer can start directly on the footing. If the footing isn't wide enough, you will need to provide an addition, as shown in Figs. 9-6 and 9-7 to insure adequate support. In some areas of the country, the building codes require the extra footing, bonded to the old footing and lower portion of the foundation wall. If the old foundation is sturdy, a steel shelf angle can be bolted to the old foundation. The anchor bolts would run all the way through the basement wall and be backed with steel plates. The steel plate must be corrosion resistant strong enough to support the weight of the brick veneer without sagging.

If you wish to use the steel plate, and have masonry or concrete basement walls, your best bet is to hire a structural engineer to devise the correct plate layout and strength. Generally, the cost of the engineer will not be as much as if you have to excavate to the footing on a full basement wall and bring the brick veneer up from there. It will be less for sure if you have to excavate 8 feet or so and then pour a large section of additional footing before erecting the veneer wall. In cases where the veneer must extend up such a distance, it seems indicated by common sense to cut masonry unit cost by using a half width cement block wall below grade. Since these half width blocks are still larger than brick in their other dimensions, the job will also go a bit faster. Check your local codes, however, to make certain this would be allowed (Fig. 9-8).

Type *M* mortar, 1 part portland cement, one-quarter part of hydrated lime and three parts of sand, should be used in below grade

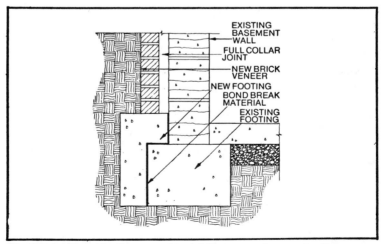

Fig. 9-7. Additional footing poured for bricking veneering construction.

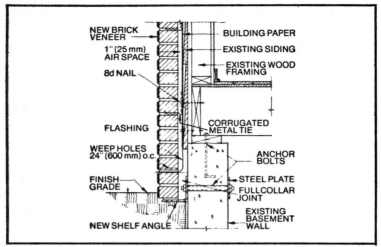

Fig. 9-8. Steel shelf angle used to support brick veneer on existing construction.

construction. Type *N* mortar, 1 part portland cement, 1 part hydrated lime and 6 parts sand, is used for above grade work.

All removable trim work around windows and doors must be moved and any fixtures such as porch railings and outdoor lights should be taken down. You'll need to make allowance for shortening porch railings and provide for slightly longer wiring for lighting fixtures. New molding, wider than the old, can be used to extend the door header, window headers and window sills to meet the new brick work. Caulking around such openings is needed and should be pressure forced into the space, using the best sealants available.

The air gap between brick veneer and existing construction is a minimum of 1 inch, just as for new construction. You will also need to add building paper over the clapboard or other existing surface to form a moisture barrier. At least fifteen pound building paper is needed for this job. As always, the air space should be kept as clean as possible. No dropped mortar should be allowed to fall onto the flashing. This would block the water and prevent proper drainage. Flashing the veneer wall is done in exactly the same manner as with new construction veneering. Weepholes are positioned at 2-foot centers in the row of head joints just above the flashing (Figs. 9-9, 9-10, 9-11, 9-12, and 9-13).

As the brick veneer wall comes up under your eaves, the overhang will be cut considerably. Before the wall comes within 3 feet of the eaves (to allow plenty of working room), the soffit should be checked and replaced if necessary. Use pressure treated lumber

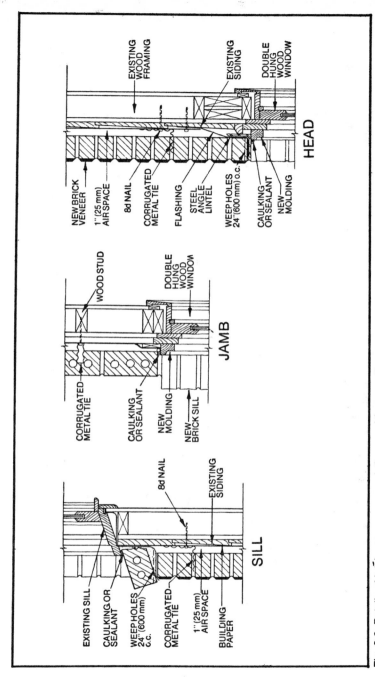

Fig. 9-9. Details of brick veneer on an existing wood frame wall.

181

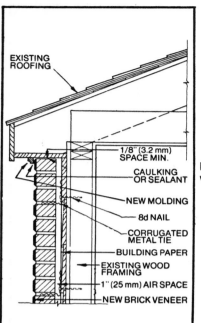

Fig. 9-10. Details for adding brick veneer to existing walls.

EXISTING ROOFING

1/8" (3.2 mm) SPACE MIN.

CAULKING OR SEALANT

NEW MOLDING

8d NAIL

CORRUGATED METAL TIE

BUILDING PAPER

EXISTING WOOD FRAMING

1" (25 mm) AIR SPACE

NEW BRICK VENEER

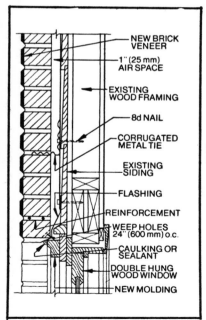

Fig. 9-11. Details for adding brick veneer.

NEW BRICK VENEER

1" (25 mm) AIR SPACE

EXISTING WOOD FRAMING

8d NAIL

CORRUGATED METAL TIE

EXISTING SIDING

FLASHING

REINFORCEMENT

WEEP HOLES 24" (600 mm) o.c.

CAULKING OR SEALANT

DOUBLE HUNG WOOD WINDOW

NEW MOLDING

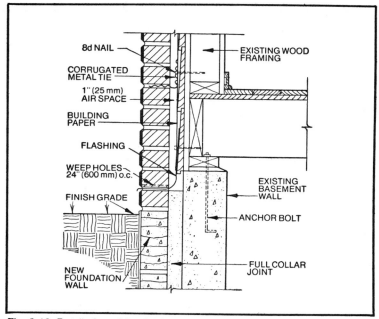

Fig. 9-12. Details for adding brick veneer.

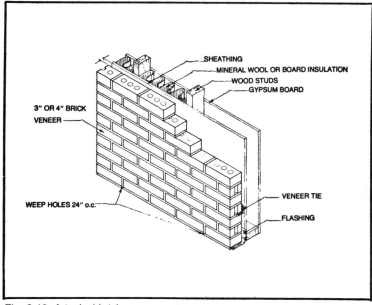

Fig. 9-13. A typical brick veneer wall.

183

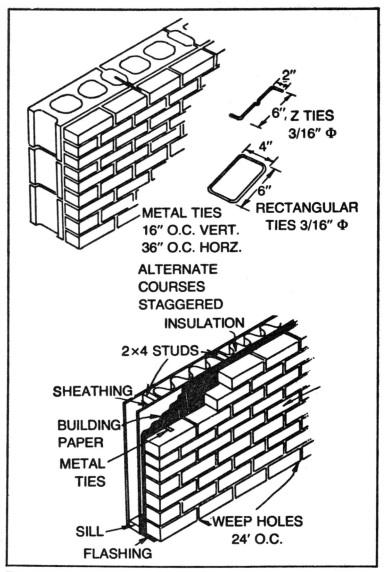

Fig. 9-14. Concrete block wall brick-veneered. Brick veneer with wood frame (bottom).

for replacement since any replacement because of rot would be very difficult with the brick veneer in the way. Detail the brick work just as with new construction and the job is complete (Fig. 9-14).

10

Reinforced Brick Masonry

There are several ways of reinforcing brick masonry to create greater resistance to tension. Reinforcements is often needed since compared to its compressive strength, the tensile strength of brick masonry is quite low. The use of special metal ties with single wythe brick walls is covered in Chapter 9. Other methods of reinforcing such walls are possible. The best and most attractive of these is probably the use of pilasters in single wythe walls over 3 feet tall. Pilasters (columns) connected to the wall at appropriate intervals can add a great deal of strength to the wall and make it much more resistant to collapse from high winds and the weight of climbing children. Generally, such pilasters have concrete footings, below the frost level. Most often the center of the column will have two or more reinforcing steel bars inserted for additional strength.

PILASTERS

Pilasters should be placed at regular intervals. Spacing depends on the expected wind load and the height of the wall. A 4- or 5-foot wall under only moderate wind load conditions (under 60 mph) might need a pilaster only every 12 feet or so. High wind conditions, or a taller wall, would require pilasters placed as close as half that distance.

When no pilasters are used, reinforced steel is used in the horizontal mortar joints. It can be used in some vertical joints, depending on the bond style used. With bond styles using staggered

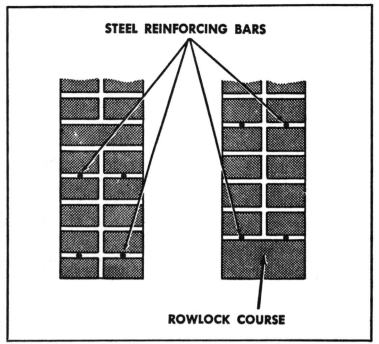

Fig. 10-1. Reinforced brick masonry, using horizontal rods.

vertical joints, in a single wythe wall, there is no way to place reinforcing steel in the vertical joints. In such cases, pilasters are still needed. Here they can be placed at wider intervals because the reinforcing steel along the horizontal joints will help tie in the wall. Metal ties are used to tie the pilasters into the wall, if the brickwork doesn't overlap.

Brick used in reinforced brick masonry must have a compressive strength of at least 2500 pounds per square inch. Type N mortar is used because of its high strength. The metal used as reinforcement is of a fairly soft grade because of the many sharp turns likely to be needed. Any wire used to tie reinforcing steel together must be at least 18-gauge soft annealed wire.

The method of laying brick for reinforced walls varies very little from standard bricklaying techniques. However, a few considerations must be kept in mind. The mortar joints must be at least one-eighth of an inch thicker than the steel used for reinforcement in order to allow one-sixteenth of an inch of mortar above and below the steel. This might mean, if large steel bars are used, that you will need to go to joints as thick as five-eighths of an inch, or even more.

HORIZONTAL BARS

In order to do any good at all, the steel must be firmly imbedded in the mortar. Horizontal bars are laid in the mortar bed and shoved down until in the proper position. There should be at least one-sixteenth of an inch of mortar under the bar. Additional mortar is then spread on top of the rods and smoothed to obtain a bed to provide a joint of the correct thickness. The brick is then laid in this bed just as it would be without the reinforcing steel bars (Fig. 10-1).

Z shaped stirrups are used to tie in vertical joints, as Fig. 10-2 shows. As you can see, laying the stirrup leg under the reinforcing bars horizontally will require a thicker than normal joint.

VERTICAL BARS

Vertical bars are placed in vertical mortar joints, held in position by wooden templates and drilled for the proper bar spacing. The brick is then laid around the vertical bars. Parallel bars must be spaced apart a minimum of one and a half times their diameter (Figs. 10-3 and 10-4).

Reinforced masonry columns and walls are useful around the home and farm. When you are erecting a brick column that is reinforced, the steel bars should have a covering of at least an inch and a half of mortar. They can be held in place with three-eighths

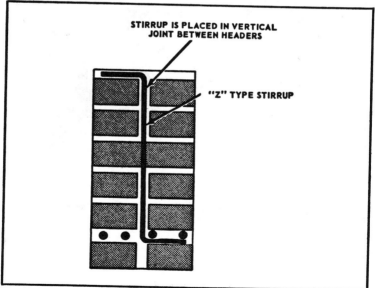

Fig. 10-2. Z stirrup used for vertical reinforcement.

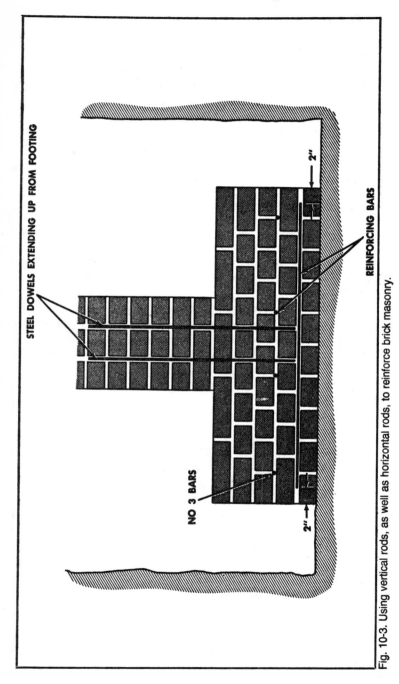

STEEL DOWELS EXTENDING UP FROM FOOTING

REINFORCING BARS

NO 3 BARS

2"

2"

Fig. 10-3. Using vertical rods, as well as horizontal rods, to reinforce brick masonry.

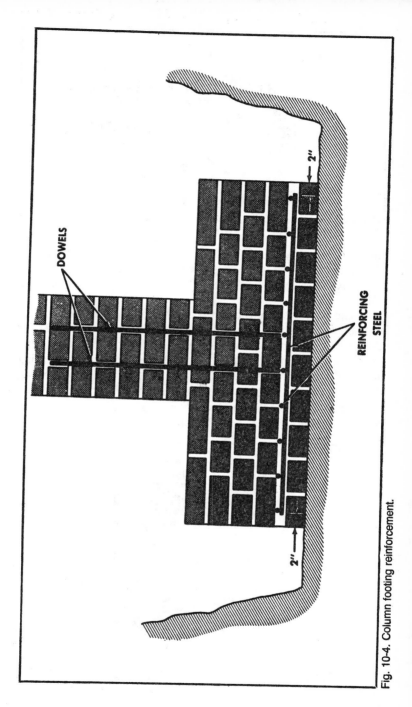

Fig. 10-4. Column footing reinforcement.

189

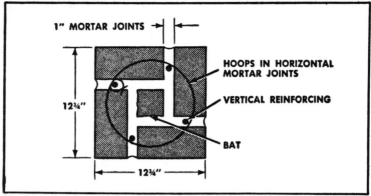

Fig. 10-5. Reinforced brick masonry column with half brick center.

inch diameter steel hoops or ties. Generally, the hoops are laid in the horizontal joints, with a single tie to one vertical bar at each joint. The steel bars are placed inside the hoop, as you see in Fig. 10-5 and 10-6.

The hoops are used at every course of brick. Once the footings are laid, with the steel rods in place, a wooden template is used over the tops of the bars. The hoops are slipped over the bars first, and one or two of them tied off to prevent interference with brick laying should the hoops slide down the bars. Then each hoop is pulled down to its course, tied to a vertical bar and embedded in mortar. For some types of columns, bats (half brick) are used in a mortar bed as column center fillers. For others, full brick is used. In all

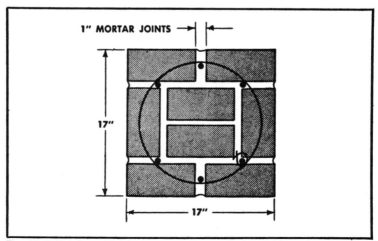

Fig. 10-6. Reinforced brick masonry column with two brick center.

190

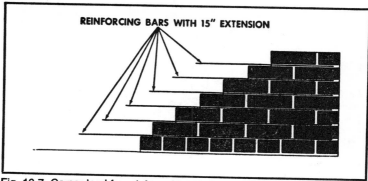

Fig. 10-7. Corner lead for reinforced brick masonry.

cases, the centers of the columns are filled with mortar and the job is done course by course as the bricks are laid.

For corner leads in walls, the bars are placed so that there is a 15-inch extension out over the ends of the leads. The bar size is the same as that used throughout the rest of the wall. Horizontal bars for the rest of the wall will overlap the corner lead bars for the 15-inch distance (Fig. 10-7).

Reinforcing steel bars for brick masonry walls will be of the same type, though usually in a smaller diameter, as that used to reinforce concrete. On the end of each bar, a number designation shows the number of eighths of an inch in diameter in the nominal diameter of the bars. Such reinforcing steel starts at three and goes up to 18. Most brick masonry will use the three, or at most the four,

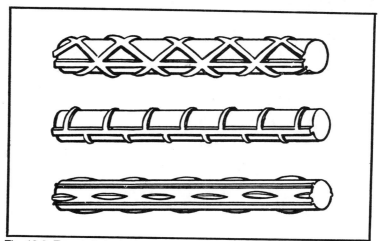

Fig. 10-8. Types of steel reinforcing bars.

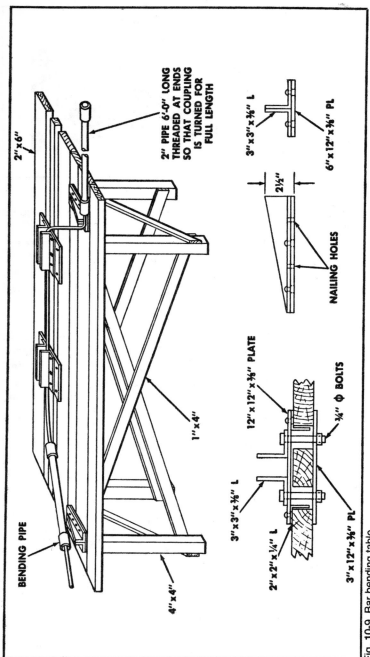

Fig. 10-9. Bar bending table.

2" PIPE 6'-0" LONG THREADED AT ENDS SO THAT COUPLING IS TURNED FOR FULL LENGTH

2" x 6"

3" x 3" x ⅜" L

6" x 12" x ⅜" PL

2½"

NAILING HOLES

1" x 4"

12" x 12" x ⅜" PLATE

¾" φ BOLTS

3" x 3" x ⅜" L

2" x 2" x ¼" L

3" x 12" x ⅜" PL

BENDING PIPE

4" x 4"

bars, since anything else forces the use of masonry joint that is too thick (Fig. 10-8).

Reinforcing steel bars are made of ductile steel to allow forming corners stirrups. You can do this work yourself using the rig shown in Fig. 10-9. Reinforcing steel should be kept free of loose rust, scale and grease.

11

Brick Paving

Brick as indoor and outdoor pavement or flooring is attractive and very durable. It's durability will exceed that of just about any other practical flooring or paving material. Variations in design are extremely varied. Many outdoor patios are simply laid in a good bed of sand, with a gravel base.

Indoors, brick can be used to accent foyers or for floor kitchens or baths (Fig. 11-1). Such floors need a bit more thought given to initial design in order to be on a level with the floors in the remainder of the home. In addition, they probably need extra support because of the increased weight. With proper initial finishing, the maintenance is so low it is often well worth the extra effort you'll have to put in to the planning to get things just right. For exterior use, special brick shapes are available and the ease of laying brick in a geometric pattern of your own design is great. This is especially true if the laying is done in sand. Mortar joints in exterior paving can be used. But this usually adds appreciably to the cost of the overall job, especially if the patio will be very large (Figs. 11-2 and 11-3).

One other consideration is cost. If your planning is good, indoor brick flooring can be laid for little more than wood covered by carpet and sometimes for less than wood floors along. This is assuming the wood is a good quality maple or oak. And of course, vinyl is cheaper than brick.

Doing the work yourself tends to cut the costs of laying a brick floor, sometimes by as much as 50 percent. Brick is often more

Fig. 11-1. A patterned and polished brick floor with mortared joints.

Fig. 11-2. A patio with a brick planter.

Fig. 11-3. Brick steps, wall and walkway.

easily laid and finished than wood flooring. Unless you are working with one of the new pre-finished wood floorings (which adds to the cost a fair amount), the brick is easier to finish to a good luster and seal. Sanding, sealing and finishing a wood floor takes a fairly skilled touch that requires quite a while to learn. On the other hand, it's possible to take a few extra bricks outdoors and practice finishing them without destroying much material at all (Fig. 11-4).

FLOORING

The advance planning of any brick floor must be complete or the final job will look terrible. In extreme cases, improper planning could cause sagging joists or other structural problems. There's not much point in installing a floor system designed to be permanently attractive and to reduce maintenance, only to have it add to problems of keeping the house standing or fail to be attractive. In other words, if the brick flooring is going in over wood joists in new construction, the distance between the joists must be reduced to provide the extra strength to support the floor's weight. Generally, this would mean moving joists that are 2 feet on center down to 15 inches on center. If the floor is especially large, move the joists to 12 inches on center. Smaller sized joists are set at 16 inches on center and could go down to 12 inches on center for moderate sized brick floors. They could be 10 inches or even 8 inches on center where the loading is especially great (Figs. 11-5, 11-6 and 11-7).

Most homes have joists running in from two sides to join at a central girder, which will then be supported on wood or steel columns. Usually those columns are spaced at least 8 feet apart. In my basement the spacing is 12 feet. In others, the girder is heavy enough that even over quite long spans no columns are needed. If the brick floor is to be a large one and runs up to or across the girder, it is a good idea to counter any possible girder sag problems by placing at least one more column under the area which will be supporting the greatest load.

In any case where you have doubts, or where your local codes require compliance have a structural engineer or architect determine the exact requirements. In many homes, the existing joist design will support considerable loads in addition to those imposed by a wood floor. In others almost any addition, other than furniture or people, could cause great structural difficulties. It's better to spend a few extra dollars to be sure and safe.

Joists under brick flooring will have solid bridging of material the same size as the joists placed every 2 feet on center.

Fig. 11-4. Imaginative use of brick creates an unusual table and a matching brick floor.

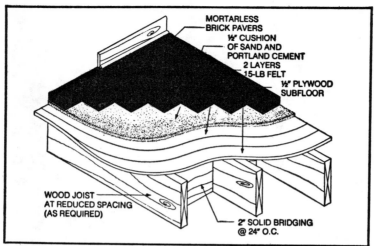

MORTARLESS
BRICK PAVERS
½" CUSHION
OF SAND AND
PORTLAND CEMENT
2 LAYERS
15-LB FELT
½" PLYWOOD
SUBFLOOR

WOOD JOIST
AT REDUCED SPACING
(AS REQUIRED)

2" SOLID BRIDGING
@ 24" O.C.

Fig. 11-5. Mortarless indoor brick floor details.

JOIST DESIGNS

A few considerations that might help you determine just what is needed in the way of improvement in either new or existing under-floor construction are in order. Almost all joist designs, even today, are set up to take the 10-pound per square foot load of a plaster ceiling. The commonly used gypsum drywall actually adds only 2 pounds per square foot to the load. This leaves an 8-pound per square foot margin to any already existing margins.

Brick used for flooring weighs approximately 10 pounds per square foot per inch of brick flooring. The pavers need not be full 2½-inch brick since new thin brick is readily available for such work. The thicknesses will vary from a one-half inch up to about 2 inches, so that the use of one-half inch brick will give you an added weight (brick alone) of about 5 pounds per square foot. Other brick flooring will be from 1⅝ to 2¼ inches thick, with a per square foot weight ranging from 16 to 23 pounds. These bricks would require reinforcement of the joist system, with as much as 100 percent additional strength required.

Normal floor deflection is acceptable at 1/360th of the span. If you don't use mortar with your brick flooring the same deflection is acceptable with brick flooring. However, you might wear out your vacuum cleaner clearing the joints. If mortar is used, the acceptable deflection drops to 1/600th of the span. Using solid bridging between the joists helps cut deflection since it forces the joists to act more as one unit.

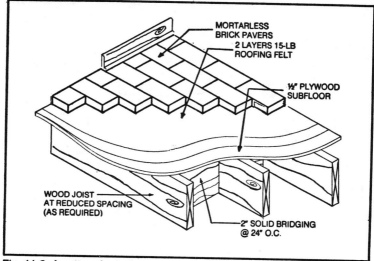

Fig. 11-6. A pattern for laying mortarless indoor brick flooring.

If the base of the floor is a concrete slab base as in a basement or for a house built on a slab, fewer problems are present when you add brick flooring. One reason for this is that there are fewer support difficulties (Figs. 11-8 and 11-9).

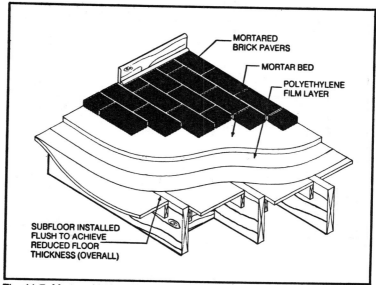

Fig. 11-7. Mortared indoor brick floors.

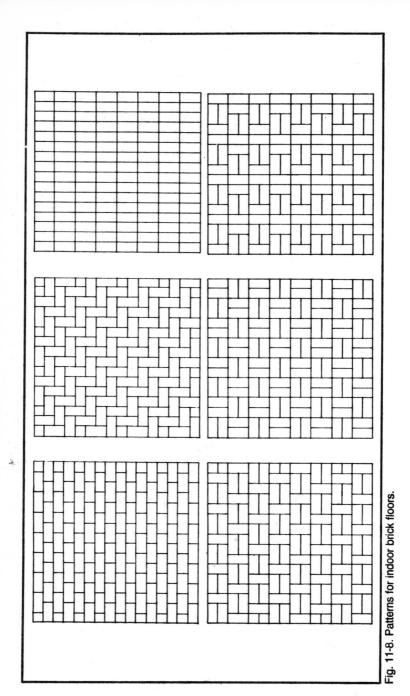

Fig. 11-8. Patterns for indoor brick floors.

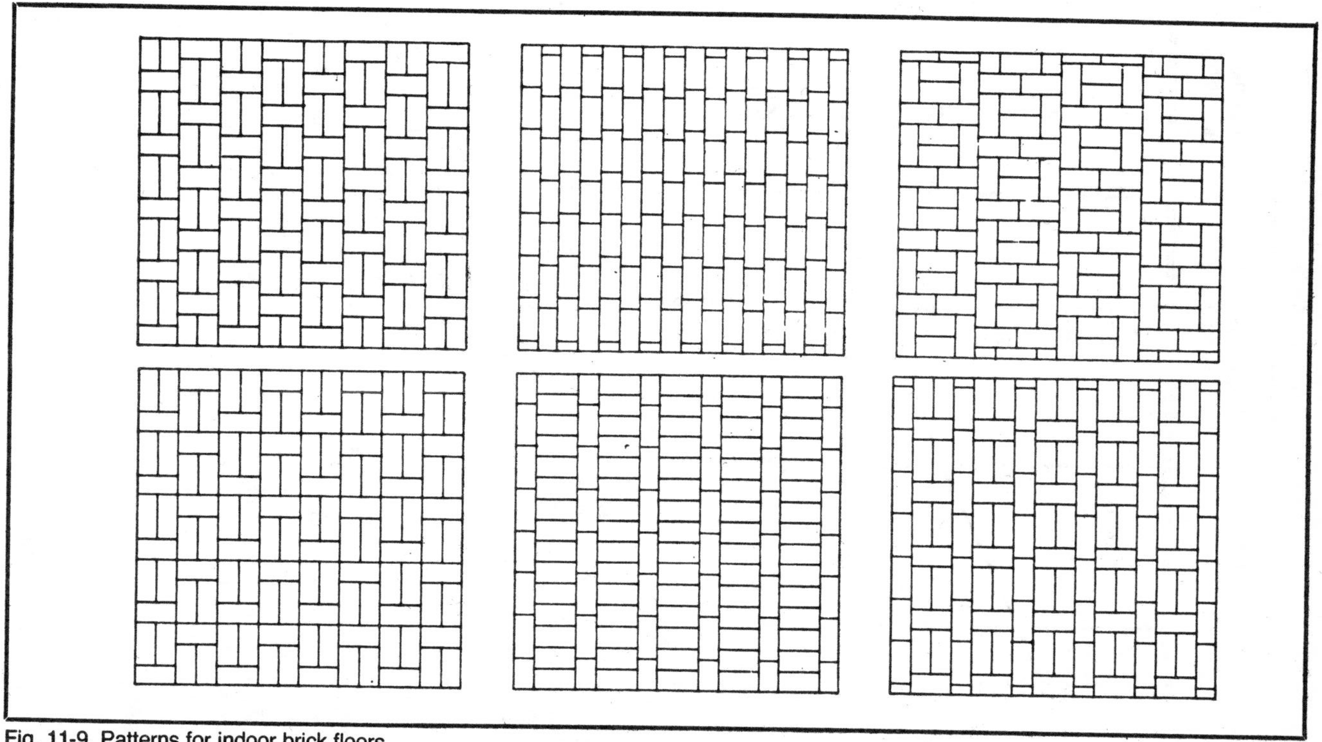

Fig. 11-9. Patterns for indoor brick floors.

203

Table 11-1. Paving Units.

Paver Face Dimensions (actual inches) w × l		Paver Face Area (in sq in.)	Paver Units (per sq ft)
4	8	32.0	4.5
3¾	8	30.0	4.8
3⅝	7⅝	27.6	5.2
3⅞	8¼	32.0	4.5
3⅞	7¾	30.0	4.8
3¾	7½	28.2	5.1
3¾	7¾	29.1	5.0
3⅝	11⅝	42.1	3.4
3⅝	8	29.0	5.0
3⅝	11¾	42.6	3.4
3 9/16	8	28.5	5.1
3½	7¾	27.1	5.3
3½	7½	26.3	5.5
3⅜	7½	25.3	5.7
4	4	16.0	9.0
6	6	36.0	4.0
7⅝	7⅝	58.1	2.5
7¾	7¾	60.1	2.4
8	8	64.0	2.3
8	16	128.0	1.1
12	12	144.0	1.0
16	16	256.0	0.6
6	6 Hexagon	31.2	4.6
8	8 Hex gon	55.4	2.6
12	12 Hexagon	124.7	1.2

NOTE: The above table does not include waste.
Allow at least 5% for waste and breakage.

For existing homes, it is often possible to insert new joists between the old joists and then to install bridging at the correct distances. If wiring or pipe placements make this difficult, the main recourse is to run 2 × 4s the full length of each joist directly against the base of the subfloor, on both sides of each joist, using at least 12-penny ring-shanked nails to attach the 2 × 4s to the original joists. As an incidental point, if you have a floor that deflects more than it should this technique on one side of each joist will remove the bounce from the floor in almost every case.

PAVER BRICKS

Working with a floor design in mind, there are quite a few types of "paver" bricks available. In making the floor design, the best idea is to work out one in which you don't need to cut any of the brick.

The floor will be more attractive with complete units. Table 11-1 shows some common paver face dimensions. There's everything from a 12-inch on-a-side hexagon, down to little 4 × 4 half bricks. A great many of these are available in thin brick, less than an inch thick. The availability of various designs and sizes varies around the country. In some areas you'll find only a few of these sizes, while in others you'll find many more.

With the thinner brick, simply recessing the subfloor by laying it between the joists instead of over them, will usually be sufficient to obtain a floor that is level, or nearly so, with the other floors in your home. With full thickness pavers, you might find it necessary to use some sort of threshold to aid in making a gradual transistion. Such transition strips should be of oak or maple and must be beveled at least on the front. If the floor doesn't take up the entire space, beveling might be needed on one or two sides as well.

PATTERNS

Once you decide on a design, there are three basic ways to lay brick over wood subflooring. Your first job is to figure out what kind of treatment the floor will have to bear in addition to withstanding the rigors of everyday traffic from however many feet. Kitchens or bathrooms where liquids are likely to be spilled or splashed are best served by brick flooring with mortared joints. In such a case, it is almost surely the best idea to use the thin brick pavers to save on weight since you will lay the bricks in a one-half inch bed of mortar (type N) over the entire floor.

It is also possible to use a portland cement-latex mortar, which will be a bit more flexible than the portland cement-lime-sand

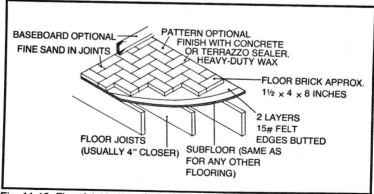

BASEBOARD OPTIONAL
FINE SAND IN JOINTS

PATTERN OPTIONAL
FINISH WITH CONCRETE
OR TERRAZZO SEALER,
HEAVY-DUTY WAX

FLOOR BRICK APPROX.
1½ × 4 × 8 INCHES

2 LAYERS
15# FELT
EDGES BUTTED

FLOOR JOISTS
(USUALLY 4" CLOSER)

SUBFLOOR (SAME AS
FOR ANY OTHER
FLOORING)

Fig. 11-10. Floor joist pattern for brick flooring.

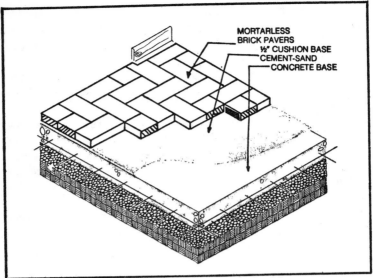

MORTARLESS
BRICK PAVERS
½" CUSHION BASE
CEMENT-SAND
CONCRETE BASE

Fig. 11-11. Mortarless brick paving laid over concrete slab.

mortars. Since there is a great deal of variation in application needs for latest mortars, depending on just who makes them, it is strongly advised that the manufacturer's directions for use be followed implicity and explicity for best performance.

A polyethelene sheet is first laid. Two layers of at least 4-mil plastic are needed, so that any slight holes won't allow water from the mortar to seep through and cause rot in the subflooring. Lay the mortar bed. Use just enough mortar at one time to accept about two dozen bricks. If the design is complex, reduce the amount of mortar bed laid at any one time. It is a good idea to lay out your theoretical design to make sure that it fits and whether your enjoyment of the pattern remains once it is down. Once the bricks are down, the joints are tooled in a conventional manner when thumbprint hard.

There are two methods of laying interior brick floors without using mortar in the joints. For one, you use full thickness paver brick and simply lay two layers of 15-pound roofing felt over at least a half inch plywood subfloor. Then place the bricks, in the selected pattern, on the felt. This is the easier of the two mortarless brick floor installations to clean, but it needs a really firm subfloor system to work (Fig. 11-10).

In most cases, a cushioned base of sand and portland cement, at least half an inch in thickness, is laid under the brick and over two layers of 15-pound roofing felt. I don't like this method.

Concrete slab bases offer the same choices for laying the brick pavers indoors (Figs. 11-11, 11-12 and 11-13).

Regardless of floor use, I would virtually always opt for the mortared brick floor. After the brick is finally cleaned, sealed and finished, the cleaning of the floor is much easier since there are no gaps for dust and grit to settle into.

The initial finishing of the brick floor will, over time, mean a great deal of difference in how easily it can be cleaned and maintained. Generally speaking, brick flooring is about as durable as a flooring can get. This assumes the proper subfloor support to keep the mortar from cracking.

You're not going to be at all happy with it if the brick dusts from normal wear and tear and the floor seems constantly dirty. As a first step to making sure the flooring is finished well, clean any stains from its surface. Make certain it is totally dry before advancing further.

In the past, a sealer was usually applied before the floor was waxed. As time went on, it was found that some sealers and waxes weren't compatible. This caused peeling and other problems. Seal a single brick with a small sample of the sealer you wish to use. Let it dry thoroughly and then use the brand of wax you intend to continue using. If there is no trouble after a week or two, then you can do the

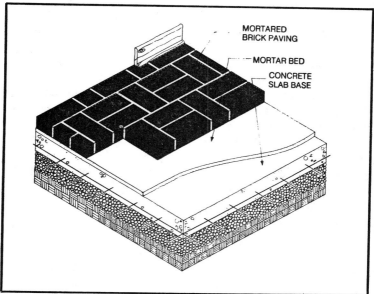

MORTARED
BRICK PAVING

MORTAR BED

CONCRETE
SLAB BASE

Fig. 11-12. Mortared brick paving laid over concrete slab.

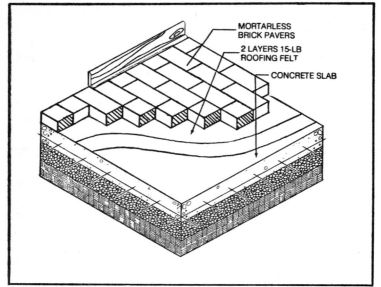

Fig. 11-13. A pattern for laying mortarless brick flooring over a concrete slab.

entire floor. For a floor with mortared joints, it will probably take this long for you to be absolutely sure the floor is completely dry.

For mortarless brick floors, there is a combination sealer/finish. This material is a liquid polymer with a plastic base and it is not permanent. About once a week you should spray a solvent solution on the floor and then use a power buffer with a steel wool pad to smooth over nicks and scratches. While a sealed and waxed floor will need to be occasionally stripped of wax and redone, weekly spraying and buffing with a machine that most homes do not have is a different problem. This is another good reason for working with mortared brick floors.

The tables (Figs. 11-1 and 11-2) will provide you with information to allow you to estimate the materials needed to do the job, along with the earlier table on paver brick sizes. In all cases, it's a good idea to add 5 percent to allow for waste because of broken brick or dripped mortar.

EXTERIOR APPLICATIONS

Mortar is almost never needed in exterior brick paving. It adds unnecessarily to the cost and complexity of the work since a concrete slab, with appropriate footings extending below the frost line is almost always needed.

Brick paving units are offered in about 40 different styles and sizes. A few manufacturers also offer special shapes for use in stairtreads and the tread bullnose. In fact, if you want a particular shape and are willing to assume the extra cost, a few manufacturers will custom make almost any shape paving brick for you. Generally, for patio and other exterior work, there is little advantage to be had in using the thinner brick styles. Stay with the standard thicknesses and you're likely to get greatly improved durability in severe use conditions such as driveways (Fig. 11-14).

Laying a mortarless brick patio, or driveway or walk, is one of the simplest of all bricklaying jobs. Select the bricks, design the patio or other area to be paved and lay the bricks on an appropriate bed for the ground conditions and that's it. There is no need to worry about mixing mortar or curing time or protecting the mortar work from drenching rains or freezing weather. A mistake is cor-

Table 11-2. Mortarless Exterior Brick Pavers.

Paver Face Dimensions (actual inches) w × l		Paver Face Area (in sq in.)	Paver Units (per sq ft)
4	8	32.0	4.5
3-¾	8	30.0	4.8
3-⅝	7-⅝	27.6	5.2
3-⅞	8-¼	32.0	4.5
3-⅞	7-¾	30.0	4.8
3-¾	7-½	28.2	5.1
3-¾	7-¾	29.1	5.0
3-⅝	11-⅝	42.1	3.4
3-⅝	8	29.0	5.0
3-⅝	11-¾	42.6	3.4
3-9/16	8	28.5	5.1
3-½	7-¾	27.1	5.3
3-½	7-½	26.3	5.5
3-⅜	7-½	25.3	5.7
4	4	16.0	9.0
6	6	36.0	4.0
7-⅝	7-⅝	58.1	2.5
7-¾	7-¾	60.1	2.4
8	8	64.0	2.3
8	16	128.0	1.1
12	12	144.0	1.0
16	16	256.0	0.6
6	6 Hexagon	31.2	4.6
8	8 Hexagon	55.4	2.6
12	12 Hexagon	124.7	1.2

Note: The above table does not include waste.
Allow at least 5% for waste and breakage.

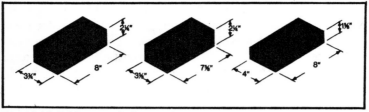

Fig. 11-14. Brick paver designs and sizes.

rected by simply lifting the incorrectly placed brick or bricks and placing them properly. The job will go very quickly, but there's no need to worry if you have to back off and let it sit—even if the wait is weeks or even months long.

Brick specifications are important for exterior paving brick since there will often be moisture present at the same time the conditions involve freezing weather. In addition, you'll need good abrasion resistance. This would almost seem to demand a dark colored hard-burned brick, since they are generally the most durable under even the severest conditions. However, you can easily select a lighter colored brick made to withstand severe weather extremes.

If you do as most people do, you'll use uncored brick and laying them with the largest dimension up. Cored brick can be used if they're placed on edge. Unless your pattern demands on edge bricks, the extra bricks needed to complete a pattern using cored brick is hardly worth the cost. It is desirable that bricks used in mortarless paving be at least twice as long as they are wide (no less than a 4 × 8 inch brick, for example), but that is not mandatory if you restrict your pattern to a one-half or one-third running bond.

Essentially, the only secret of a good mortarless paving job with brick is proper preparation of the subsurface. Pay a lot of attention to drainage. Slope the patio or other paved area away from the house. A slope of one-eighth of an inch to one-quarter of an inch will be enough for small areas. If you live in an extremely wet area, or an area with poor percolation features, greater attention to drainage detail will be needed in most areas today, before a house can be built on a lot, a percolation test has to be run. If you're curious about the soil percolation rate around your home, and the home is relatively new, check with your local building inspector to see if his office has the figures. In most cases, a layer of clean gravel—not crushed stone—carefully sloped for drainage will solve the problem. The brick pavers can then be set directly over the gravel (Fig. 11-15).

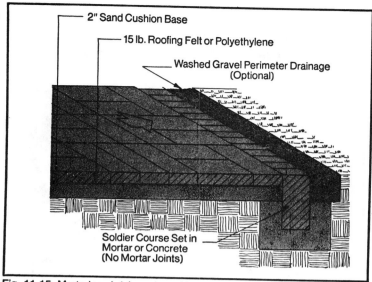

Fig. 11-15. Mortarless brick paving with a cushion of sand for a base.

In more extreme cases, you will probably need a deeper layer of gravel at the edge or at the wettest spot of the paved area, and a clay drain tile run through the deeper trough filled with gravel. In such cases, the minimum base will be about 4 inches of washed

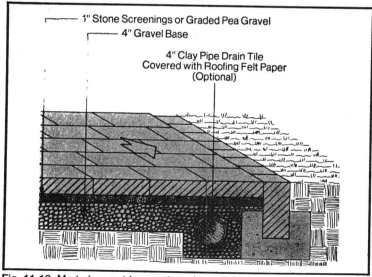

Fig. 11-16. Mortarless outdoor paving where drainage needs are greater.

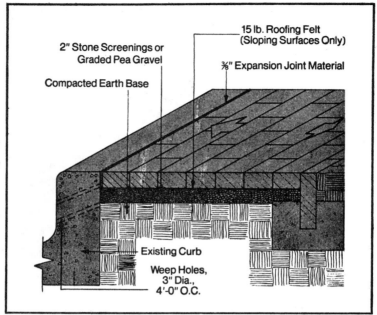

2" Stone Screenings or Graded Pea Gravel

Compacted Earth Base

15 lb. Roofing Felt (Sloping Surfaces Only)

⅜" Expansion Joint Material

Existing Curb

Weep Holes, 3" Dia., 4'-0" O.C.

Fig. 11-17. One method for preventing edge movement in driveways and walkways.

gravel, with anywhere from 8 inches to 24 inches of the same kind of gravel surrounding a 4-inch clay drain tile leading away from the paved area. On top of the 4 inches of washed gravel, you place about 1 inch of stone screenings or pea gravel to provide a base for the brick pavers. Both layers of gravel should be carefully tamped (Fig. 11-16).

For the lucky people without drainage problems, the brick pavers can be set directly on 15-pound roofing felt laid over a 2-inch sand base, carefully tamped and leveled.

For driveways, the Brick Institute of America states that the gravel base can be reduced as the thickness of the paving unit increases. If you use a 3½-inch cored brick, standing on edge, you can reduce the gravel depth to 2 inches from the 5-inch gravel base needed for 2¼-inch brick.

While BIA recommends road construction capability compactors for the crushed stone, almost all such work can be done with hand tamping equipment or by renting the services of a local paving contractor with a small roller for an hour or two. Hand tamping should be allowed to settle for at least two weeks before any brick is laid.

Plastic membranes or roofing paper can be used under the brick. Such a procedure makes laying the brick exceptionally easy since the surface is smooth. This method tends to prevent the rise of moisture from the ground into the brick, with the possibilities that brings of staining the bricks. Still, such membranes are never essential.

Mortarless brick driveways have a disturbing tendency for the edge paving units to shift under the weight of passing vehicles. Some walkways could suffer the same fate under vigorous foot traffic. The fastest and best cure for such shifting is to simply set a soldier course of bricks in a firm mortar bed, at least a foot deep. The soldier course, as you remember, is a brick standing on end. Install the edging before placing the pavers and it will serve as a guide for the entire rest of the job (Figs. 11-17, 11-18, 11-19).

Once the mortarless brick pavers are in place, sift sand over their surfaces and then simply sweep it so that is fills in the cracks between bricks. In deference to those who think a mortared joint looks a bit better but who don't wish to go to the trouble and expense of laying a slab, the sweeping can consist of three parts of mason's sand and one part of portland cement. Dampen the mixture by giving it a light spray with your garden hose (Table 11-2).

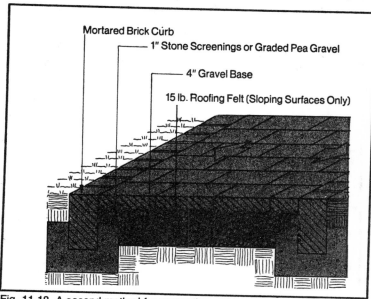

Fig. 11-18. A second method for preventing edge movement in driveways and walkways.

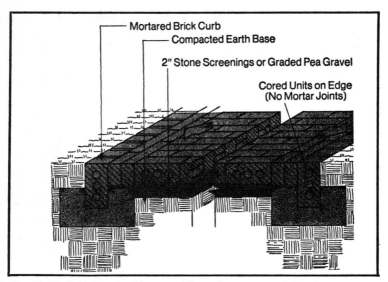

Fig. 11-19. A third method for preventing edge movement in driveways and walkways.

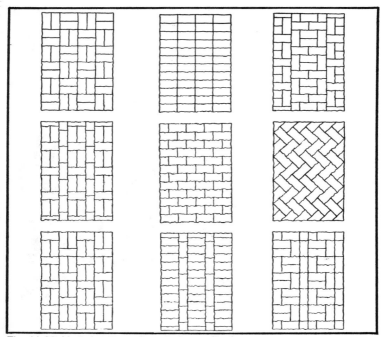

Fig. 11-20. Varied patterns for exterior application.

Mortared exterior paving must be set on a slab in most areas of the country since the freeze/thaw cycle will cause cracking of the joints if the base isn't solid and almost totally immovable. The same type of bricks will be used and the mortar will be type *M*. Use 1 part portland cement, one-quarter part lime and 3 parts sand, by volume. The bricks are set in a 1-inch bed of mortar and the entire job must be kept damp for at least three days to assure proper curing.

Personally, I can see little or no advantage to the mortared exterior paving in any context. I believe that it does little but add expense and work to what is, primarily, the simplest and most basic bricklaying job of all. Mortarless outdoor paving, with sand filled joints, is exceptionally attractive (Fig. 11-20) and extremely durable. It is also easy to repair if you accidentally drop a heavy object and crack a brick. Just lift out the cracked brick, clean away the sand, set in a new brick, sweep sand back into the joints and you're done.

The Brick System House

A few years ago, the Brick Institute of America commissioned the design of a quite interesting house. Several stipulations were made to make the project a practical one for most building contractors. I'm a firm believer in doing it yourself and I do think the average competent and healthy person has the capability of building his or her own home. However, this one is, at first look, beyond the capabilities of a novice bricklayer. Still, it should be of interest to the more advanced bricklayer. It might even be a possibility for those planning a home and hoping to hire a contractor to build it (Fig. 12-1).

The primary interest of BIA was to prove that brick is as economical as most other building materials, that the work should be done in about the same time or even less than could conventional building, and that there was no need for specialized craftsmen or special supervision. In essence, a firm of architects was hired and the plans turned over to a builder. The house was sold well before completion although many frame homes in the area remained unsold for some time after completion.

The BIA stipulated to the architects—Strang, Childers & Downham, AIA, of Annandale, Virginia—that the project had to built without special contractors and that it had to be competitive in price with houses of similar space, equipment, land and floor area. The house had to be built near a major metropolitan housing area near similar houses (built of different materials). The house was

Fig. 12-1. The Brick System House.

built by a currently active homebuilder and materials were similiar to those readily available throughout the country. One further stipulation was that the experimental features, the brick exterior wall and interior finish, were to be observed and incorporated, but they were to have no special supervision or inspection during construction.

The actual construction took about five and a half months. Part of this time was during winter, not at all out of line with other home building procedures.

The Brick System House has three bedrooms, a living room, a dining room and a full basement with a roughed in third bath. This demonstration model added a brick paved breezeway, a carport, decorative brick screen walls, brick columns, a brick patio and a barbeque. Total living space is 1260 square feet, and the lot is about a quarter of an acre (Fig. 12-1).

EXTERIOR WALL SYSTEM

An experimental feature of the house is the brick exterior wall system (Figs. 12-3 and 12-4). Essentially, it is a single wythe wall of 4-inch brick, with 4-inch thick pilasters placed every 4 feet on center for stiffening. Load resistance is estimated at 25 pounds per square foot. There is no interior stud wall as there is in brick veneer construction, so the pilasters serve as a base for both the 1½-inch rigid styrofoam insulation (adding a 4-inch dead air space between pilasters) and for the interior drywall finish walls. Moisture penetration is prevented with the use of a cement parging on the backs of the wall bricks. The three-eighths of an inch parged coating is then coated with asphalt for further protection. The foundation wall for the full basement is constructed of concrete block and has a one-half inch parged cement coating. Weepholes are placed above the flashing at the base of the brick wall on 4-foot centers for drainage. This distance could be reduced to 2 feet in extreme moisture conditions.

The exterior brick wall has metal Z-ties to the pilasters in every fourth course of brick.

FLOORING

Flooring is pretty much standard, with 2 × 10 joists laid on brick sills that are tied in with metal straps. This is then covered with five-eighth inch tongue and groove plywood for the subfloor. The finish floor can be of the builder's selection.

The experimental wall was designed to meet all model building codes for brick masonry and should meet just about all reasonable local masonry and should meet just about all reasonable local codes.

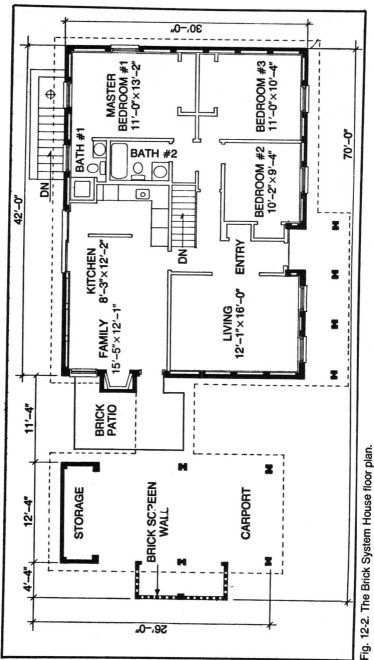

Fig. 12-2. The Brick System House floor plan.

219

While still staying within the projected price range, the Brick System House demonstration model was supplied with a number of amenities not often found in what are generally classified as development houses. These include the all-brick fireplace, a brick paved patio, screen walls of brick near the carport and the decorative columns supporting the carport and the front porch roof.

If this house interests you, plans are available from the Brick Institute of America, 1750 Old Meadow Road, McClean, VA 22101. The first set of plans is $25.00 and additional sets bought at the same time are $7.50. The plans not only include specifications for the brickwork, but cover all other requirements such as heating and electrical details.

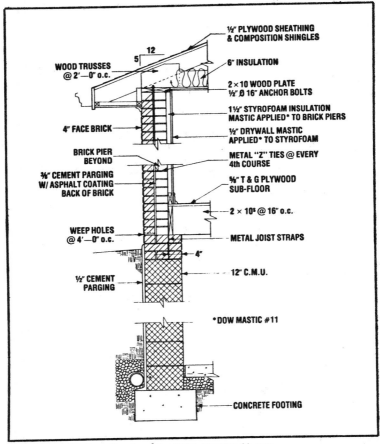

Fig. 12-3. A Wall section from the Brick System House.

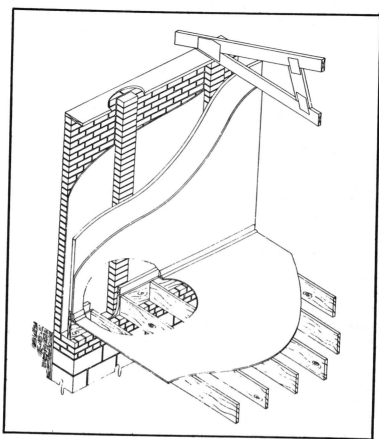

Fig. 12-4. Flooring and joist details.

The Brick System House was considered successful for several reasons. It was built, as intended, by face veneer bricklayers and it was built and sold for just about the same price as neighboring houses, some of which lacked several or all of the extras and none of which were of brick masonry construction. While working on the demonstration model, the builder received several requests to build similar houses.

There are really only two innovations with this house, but those successfully challenged two long held assumptions in the construction industry. First, it was proven that an all brick house is not too difficult for ordinary masons to build. Second, it was also proven that brick in this sort of construction can be as cheap as conventional frame construction of good quality.

Fireplaces And Chimneys

Over the years, fireplace construction has been more than a little hit and miss. During the 1700s, Count Rumford formulated the design shape that even today produces the greatest amount of heat for the amount of wood used. With energy costs ever more expensive, and likely to continue to increase as the availability of fossil fuels decreases, the burning of wood as a main or supplementary heat source has taken off like a rocket. Last year something like a million wood stoves were sold in the United States and at least that many were sold in the first three-quarters of the present year. Almost everyone I know is heating at least a portion of their homes with wood. I have been using two wood stoves as a complete heat source for two winters now.

Unfortunately, conventional fireplaces often prove to be extreme heat washers as they tend to draw air heated by the furnace up the chimney and to use heated home air for combustion. Also, a fireplace is difficult to shut down at night, since the house will fill with smoke when the damper is closed to prevent night time heat loss up the chimney. It has been said that the average fireplace adds 10 percent or a bit more to the cost of heating a house. That's not exactly a fine idea with the price of fuel oil, gas or electricity today.

MODIFICATIONS

Fortunately, some modifications to existing fireplaces can produce much better results and proper design in a new fireplace can

Fig 13-1. A conventional masonry fireplace, with hood.

add dramatically to the heat producing efficiency. Although no fireplace will ever approach the efficiency of today's top woodstoves, for most of us the fireplace adds a romance to the house not available with a woodstove. Basking in front of a blazing fire with cognac in one hand and your other arm draped around the mate of your choice is one of the greater pleasures in the modern world (Fig. 13-1 and 13-2).

For existing fireplaces, the basic idea is to provide some sort of entryway for outside air to aid combustion, so that warmed house air is not wasted. The modifications shown in Fig. 13-3 to allow that passage of exterior air to the fire show how things can be done if the fireplace is being designed ahead of time. The intake is located on an outside wall or at the back of the fireplace. The louver used should be screened to keep out insects in warm weather and should have closable louvers for days when the fireplace isn't in use. Otherwise the cold air will be drawn into the house when it isn't needed, again reducing fireplace efficiency. It's best if the louver can be operated from inside the house, but this is not always possible and not essential.

The passageway for average size fireplaces will need a cross sectional area of 55 square inches. The passageway can be built, as shown, in the base of the fireplace or it can be run down between the joists of the house and then to the intake. No matter how the passageway is built, it needs to be well insulated so that cold air is not transferred to the interior of the home.

The inlet brings the outside air into the firebox and requires a damper for air volume and direction control. The damper should be located at the front of the firebox and in front of the log grate. Most such inlets will be about 4 ½ × 13 inches, though the damper you use might force you to go a bit larger or a bit smaller. The air intake pit is located directly below the inlet, with the same dimensions as the inlet, and should be approximately 13 inches deep for best draft.

There is a twofold benefit in using this modified fireplace design. First, the outside air is used to generate the draft and keep a good fire burning. This can be of particular importance in modern, tightly caulked and heavily insulated homes where it is often necessary to crack a window to get enough air flow to allow proper combustion in a fireplace. It also helps cut down on problems of waste in older homes. Second, in homes that are not so tightly sealed, the infiltration of air to feed the fire is cut down so that less air needs be heated. Also, the drafty feeling of infiltrating air is cut down.

Fig. 13-2. With care, even the amateur brick mason can produce a structure such as this, although it certainly is not a total beginner's project.

Fig. 13-3. A conventional fireplace, with exterior air supply.

When the fireplace isn't in use or must be shut down for the night, you should make sure of several things. First, the fireplace damper should fit tightly for those times when the fireplace is totally shut down. A glass fire screen should also be used since this allows you to shut the fireplace down at night and prevent furnace-heated air from being drawn up the chimney after the fire dies out. The glass firescreen should fit the fireplace opening tightly. This can be a

Table 13-1. Conventional Fireplace Dimensions in Inches.

Finished Fireplace Opening							Rough Brick Work			Flue Sizes [a]			Steel Angle [b]
A	B	C	D	E	F	G	H	I	J	K	L	M	N
24	24	16	11	14	18	8¾	32	18½	19	10	8 × 12		A-36
26	24	16	13	14	18	8¾	34	18½	21	11	8 × 12		A-36
28	24	16	15	14	18	8¾	36	18½	21	12	8 × 12		A-36
30	29	16	17	14	23	8¾	38	18½	24	13	12 × 12		A-42
32	29	16	19	14	23	8¾	40	18½	24	14	12 × 12		A-42
36	29	16	23	14	23	8¾	44	18½	27	16	12 × 12		A-48
40	29	16	27	14	23	8¾	48	18½	29	16	12 × 16		A-48
42	32	16	29	14	26	8¾	50	18½	32	17	12 × 16		B-54
48	32	18	33	14	26	8¾	56	20½	37	20	16 × 16		B-60
54	37	20	37	16	29	13	68	22½	45	26	16 × 16		B-72
60	37	22	42	16	29	13	72	24½	45	26	16 × 20		B-72
60	40	22	42	16	31	13	72	24½	45	26	16 × 20		B-72
72	40	22	54	16	31	13	84	24½	56	32	20 × 20		C-84
84	40	24	64	20	28	13	96	26½	61	36	20 × 24		C-96
96	40	24	76	20	28	13	108	26½	75	42	20 × 24		C-108

[a] Flue sizes conform to modular dimensional system.

[b] Angle sizes: A—3 × 3 × 3/16 in.; B—3½ × 3 × ½ in.; C—5 × 3½ × 5/16 in.

[c] This dimension is listed to provide a minimum thickness of the fireback.

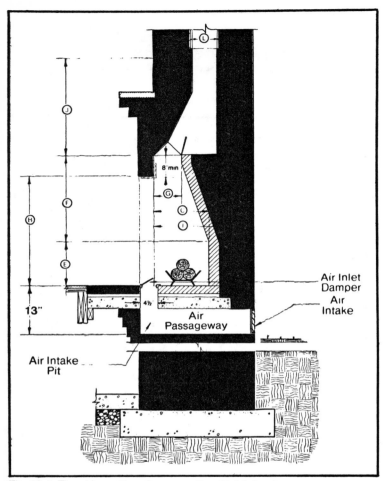

Fig. 13-4. The side section of a conventional brick fireplace.

chore with older fireplaces made of materials other than brick, but it is well worth the effort involved. The firescreen also should be the type with a row of louvers or holes along the bottom which can be tightly closed. The best type of firescreen does not have any venting at all since such vents are not needed with modified fireplaces.

For fireplaces already installed, the modification for the intake of exterior air could prove to be a somewhat complex chore. You might find it impossible to put an air passageway through the base of the fireplace, especially an all masonry fireplace built on its own footing. Such a base would be massive. If the fireplace has an ashpit, you could run a duct from that to the exterior wall and replace the ashpit door in the hearth with a damper. Or simply prop the ashpit door open to save some time and money. In any case, it's unlikely the dimensions of the ashpit duct will provide enough exterior air for perfect combustion. In such cases, you should use a glass firescreen with adjustable louvers to allow the use of some interior air if it becomes necessary.

Table 13-1 shows the needed dimensions for a conventional fireplace. Figures 13-4, 13-5, and 13-6 indicate the application of those figures at the various points in the fireplace. While correct flue sizes are imperative to correct fireplace functioning, the charts do not include a couple of figures which are just as important. If flue height is not correct, the fireplace will smoke. If obstructions are too near the top of the chimney or the top of the chimney is not far enough above the roof line, the fireplace will not draw properly. The flue must rise at least 14 feet from the firebox's top opening, or lintel. Then the chimney, on peaked roofs, must rise at least 2 feet higher than any obstruction such as the roof peak or a tree within 10 feet. On flat roofs, the chimney must rise 3 feet instead of 2 feet. These figures are true for all chimneys, whether they are made of masonry, metal or other materials and for either a fireplace or for a wood stove.

THE RUMFORD FIREPLACE

The fireplace designed by Count Rumford differs quite a bit from the conventional fireplace. The major differences are found in

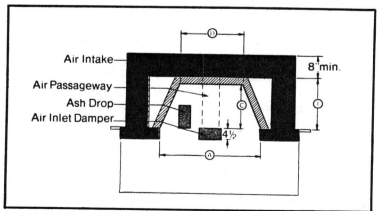

Fig. 13-5. Top view of a conventional brick fireplace.

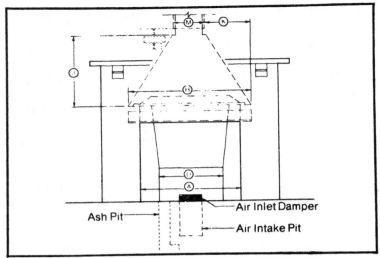

Fig. 13-6. Front view of a conventional brick fireplace.

the firebox. The Rumford fireplace features a shallow firebox, with obliquely flared sides and back so that more heat is radiated into the room than is with the deeper conventional firebox. If the Rumford design is incorporated with the exterior air source and a tight glass screen efficiency will be quite a lot greater than with a conventional design and no more difficult to build.

Referring to Figs. 13-7 and Fig. 13-8, the Rumford firebox width (D) must equal the depth (C) and the vertical portion (E) must equal the width. The thickness of the firebox (I minus C) should be at least 2¼ inches, while the area of the fireplace opening (A × B) must not exceed 10 times the flue opening area. The width of the fireplace opening (A) and its height (B) should each be from two to three times the depth of the firebox (C). Opening height (B) should be no larger than the width (A). The throat of the fireplace (G) should be no less than 3 inches nor more than 4 inches and the centerline of the throat must align with the centerline of the firebox base. The smokeshelf (R) will be 4 inches wide, regardless of other dimensions. The lintel (0) is at least four inches wide but no more than 5 inches wide. The vertical distance from throat (P) to lintel must be at least 12 inches. You are required to use a flat plate damper and it must open towards the smokeshelf.

Combine these features with exterior air for combustion and a glass firescreen and you'll have the most efficient fireplace of total masonry construction possible.

Actual construction of the fireplace follows the basic rules of bricklaying. A masonry fireplace is built on its own footing, which must be fairly massive to support the weight of the fireplace and chimney. If you are adding to an existing structure, the entire footing is poured outside the foundation wall, going down below the frost line. In all fireplace footing construction, the footing is free floating. That is, it abuts the house foundation or footing, but it is not joined to it.

When constructing the hearth, it should extend at least 2 feet in front of the fireplace opening and a minimum of a foot to each side of the fireplace opening. The hearth must have a base of at least 4

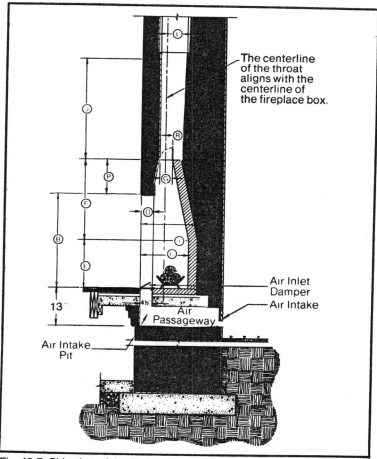

The centerline of the throat aligns with the centerline of the fireplace box.

Air Inlet Damper
Air Intake
Air Passageway
Air Intake Pit
13"

Fig. 13-7. Side view of the Rumford fireplace.

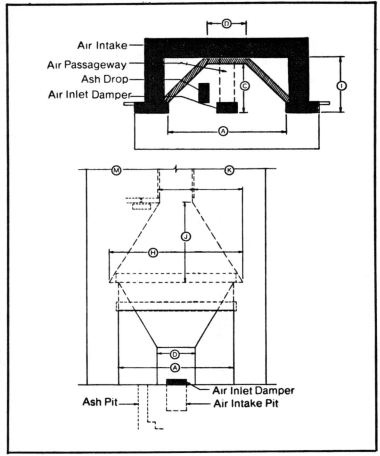

Air Intake

Air Passageway

Ash Drop

Air Inlet Damper

Ash Pit

Air Inlet Damper

Air Intake Pit

Fig. 13-8. Top and front views of the Rumford fireplace.

inches. This is usually poured concrete, but it often can be 4-inch brick laid on its side in a good mortar bed. Usually it is best to make the hearth flush with the floor since this makes the fireplace easier to clean. A flush hearth is also more easily installed than cantilevered and raised hearths. The firebox floor is also called the hearth and should not be built of regular brick or concrete.

The back and sides of any masonry fireplace built today will need to be solid masonry at least 8 inches thick and lined with firebrick in order to pass building codes. Actually, the codes also approve the metal liners so popular in many areas. If you're building a Rumford style fireplace, (I recommend the Rumford over the conventional) you would probably have to have a metal insert custom made. I have never seen one advertised. A few codes will allow the elimination of firebrick or other linings if you use 12-inch thick masonry. But that is not practical since the standard masonry units will crumble fairly rapidly under the heat conditions in a fireplace.

Your chimney will need to be at least 4 inches thick, with fireclay flue liner to meet codes. The flue liner is joined with fireclay mortar just as are the firebricks in the firebox. Chimney offsets— deflection of the flue from the vertical—should be avoided whenever possible as they add to creosote accumulation. In no case should the offset exceed 30 degrees. For an offset of that size, you would be wise to increase the flue cross sectional area at least 20 percent, with a 10 percent increase if the offset is 15 degrees.

As you reach the top, the final flue liner should extend at least 4 inches from the chimney's top surface (surface of the brick). A taper or bevel of mortar is laid to within 2 inches of the top of this last flue liner in order to direct water off the top of the masonry units. This also provides a very slight increase in the draft.

TESTING A FIREPLACE

Now comes the final test. You have got the fireplace in and the chimney up. You can't build a fire until the chimney is well cured, usually at least 2 weeks after the last mortar is placed. However, you should test the unit for smoke leakage. Essentially, you build a smudge fire that puts out very little heat but a lot of smoke. A smudge pot, if you can locate one, is fine. A small hibachi or metal barbecue could be used. Build a fire using a few sheets of paper and a few handfuls of damp straw, or a bit of old roofing felt or some other such material. While the fire is burning place a batch of wet newspapers over the chimney opening. Even small leaks will be very obvious as the murky smoke is forced back down the flue. That murkey smoke is the reason you only want to build a small smudge fire. A large fire could easily fill the house with smoke. Any leaks should be attended to immediately.

Let the fireplace and chimney cure for two to three weeks before laying the first real fire. For another two to three weeks, keep the fires small. At this point, a roaring blaze could cause cracks in the mortar. That is unnecessary and it could be dangerous.

CHIMNEY FLUES

Each fireplace or wood stove *must* have its own flue. Though a separate chimney may not be needed, an individual flue is essential

Table 13-2. Fireplace Flue Sizes (In Inches).

Fireplace opening			Standard rectangular flue, outside dimensions	Round flue, inside diameter
width	height	depth		
28	24	16-18	8½ × 8½	10
30-32	28	18	8½ × 13	10
36	28	18	8½ × 13	12
42	28	18	8½ × 18	12
48	32	18-20	13 × 13	15
54	36	20	13 × 18	15
60	36	22	18 × 18	18

to proper draft, which is essential to proper burning. Tables 13-2 and 13-3 list flue sizes.

Recommended flue sizes are the same for fireplaces with openings of 2400 to 3000 square inches. If your fireplace is larger than that, and your chimney is about 15 feet high or more (from hearth to chimney top), the flue opening, or flue area, should be about 1/10 the area of the fireplace opening. The flue area should be about the area of the fireplace opening if the chimney is under 15 feet high. Select the nearest commercially available size and use that for chimney construction. All the sizes listed above are commercially available.

If a single flue won't handle your fireplace, you can install two flues in a single chimney. It's more work and should only be done if there is no other way to get a flue of sufficient size. The flues, if doubled, must receive the same treatment as any separate flues. There must be at least 4 inches separating them, and one should be 4 inches higher than the other to prevent downdrafts. If one flue of a double set causes a downdraft into the other, the fireplace will smoke almost as badly as if it had a single flue receiving a downdraft.

It's also possible to build a flue too large. Too large a flue area is almost as bad as one that's too small. Do not exceed the recommended sizes. If the chimney height is exceptional, then a slight reduction in flue size can help to keep the draft within limits.

MASONRY CHIMNEYS

As stated earlier, all chimney tops should rise at least 2 feet above the roof ridge on peaked roofs. On flat roofs, the chimney should rise a minimum of 3 feet above the surface. This distance can reduction in flue size can help to keep the draft within limits.

Table 13-3. Alternate Flue Sizes.

Area of fireplace opening in square inches	Standard rectangular flue, outside dimensions	Round flue inside diameter
1000	8½ × 18	12
1200	8½ × 18	12
1400	13 × 13	12
1600	13 × 13	15
1800	13 × 18	15
2200	13 × 18	15
2400	18 × 18	18

be increased, but under no circumstances should it *ever* be reduced. In the case of peaked roofs, a chimney should extend at least 2 feet above the highest point within 10 feet, which means, of course, that with proper placement the chimney need not always rise 2 feet above the *ridge peak*. But if there are large trees near the house, you may have to increase the height of the chimney to prevent downdrafts. Disturbed air currents around those trees may knock smoke back down the flue and into the room (Fig. 13-9). In some cases, you'll almost certainly find it simpler and easier to remove obstructing trees or branches rather than run a silly looking chimney a couple of dozen feet over the roof line of your house. Nearby hills and tall buildings can cause similar problems.

You'll need a reasonably careful study to figure final chimney height. In almost all cases, whether you're putting up a masonry or a prefabricated chimney, a trip to the roof to take measurements and a careful look at surrounding buildings, hills, and trees can save a lot of later grief.

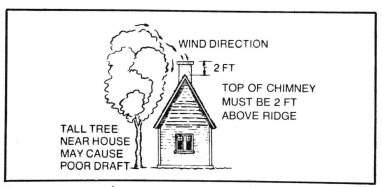

WIND DIRECTION

2 FT

TOP OF CHIMNEY MUST BE 2 FT ABOVE RIDGE

TALL TREE NEAR HOUSE MAY CAUSE POOR DRAFT

Fig. 13-9. Tall obstructions near a house can cause chimney draft problems.

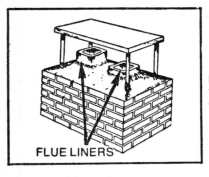

Fig. 13-10. A pair of flues set in mortar atop a capped chimney (courtesy of Majestic).

FLUE LINERS

You can use a straightedge, tacked to the side of the house where the chimney will rise, as a guide. Tie the end of a 10 foot length of mason's cord to the straightedge and locate the highest point on the roof within the reach of the cord. Then measure from the highest point 2 feet upward. The distance from 2 feet above the highest point to the throat of the fireplace is the height of your chimney. The throat of the fireplace is where the flue starts.

The exterior of the masonry chimney can be built with several materials in several ways. Cement blocks still are the cheapest (and the ugliest when left unfinished). Bricks with a fireclay flue liner offer sufficient thickness (4 inches) but take time to lay. Because of their small size, common bricks are time consuming to lay but are also quite attractive and appeal to a good many people. Cement blocks offer the greater construction ease, but usually require some sort of covering, whether brick veneer or some special painting or stucco.

As the flue liner is set in mortar and built on up (Fig. 13-10), the exterior of the chimney should be maintained just one or two tiles behind. This makes for greater working strength for the flue liner and also makes the job look a bit more orderly at the end of each working day. Flue liner by itself doesn't supply much strength, so the surrounding masonry keeps you from knocking it over as quickly as it goes up. Try to bring the course of blocks, stone, or brick almost to the top of the final flue tile installed on any given day.

In most cases, you'll need scaffolding to do the masonry work on the chimney since you'll need to have quite a lot of mortar, bricks and tools on hand at working height. A chimney maker trying to work from a ladder is giving up a lot of just to save a few bucks rental fee. The scaffolding will allow you to work all around the chimney and it will only need to be raised after every 4 or 5 feet of chimney is laid.

Flues without flue liners may have very rough interior mortar seams. Such rough seams interfere with air flow. One worker I know has a good way to cut down on seam roughness. First, he takes a heavy cloth sack and stuffs it full of wood shavings or straw.

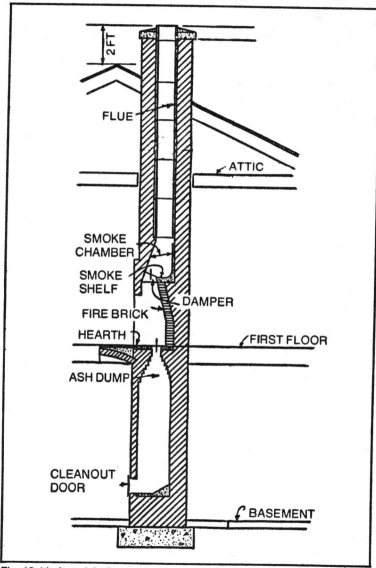

Fig. 13-11. A straight flue in a masonry chimney.

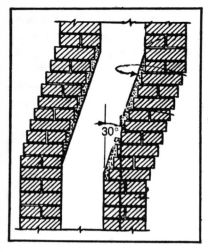

Fig. 13-12. A flue offset of 30 degrees in a masonry chimney.

Then he ties the neck securely and fits the bag snugly into the flue just above the fireplace. A heavy piece of twine leads from the bag on up to his chimney working platform. After he finishes a course or two of flue tile, he pulls the bag up. The snug-fitting bag slides up the flue, smoothing down mortar joints as it goes. This technique means that little or nothing can work its way down the flue and mess up the smoke shelf of the fireplace. The bag should be twisted as it's pulled up for best results.

The simplest type of flue installation is the straight flue (Fig. 13-11). This kind of flue runs straight up from the fireplace, leaving little space for debris to collect and presenting the first time builder with the fewest problems of any masonry chimney construction. Offsets in any direction can be built into any chimney, but they can add to soot accumulation, resin and creosote buildup, and draft problems. They are also more complicated to build. As stated earlier, in no circumstances should any flue offset exceed 30 degrees (Fig. 13-12). For a chimney with a 30 degree offset, it's a good idea to increase the flue size (cross-sectional area) by 15 or 20 percent; a 5 or 10 percent increase in flue size is helpful with a 15 degree offset.

The final step in masonry chimney construction is to make sure that the final fire clay flue liner section is at least 4 inches from the chimney's exterior surface. A taper or bevel of concrete should be laid to within 2 inches of the liner top, as shown in Fig. 13-10. This bevel helps rain water to run off and will provide some help in forming an updraft to help prevent smoking of the fireplace.

Fig. 13-13. A Majestic fireplace with a prefabricated chimney.

You must make a careful check to ensure that the completed chimney will not leak smoke to the inside of the building. Such smoke leaks can be dangerous. One problem with any masonry fireplace and chimney is the length of time you have to wait before using the installation, for the mortar must have time to cure properly before much heat is applied. Still, you should make a leak test as soon as the installation is completed.

PREFABRICATED CHIMNEYS

Most factory-built fireplaces, as well as most wood stoves, come with prefabricated low- or no-clearance insulated chimney sections with built-in flues (Fig. 13-13). These 2- and 3-foot chimney sections make chimney installation a breeze. All the leading chimney systems come with twist or snap locks to give a tight seal between sections. Standard straight sections are available in 2- and 3-foot lengths and weigh less than 30 pounds apiece, so chimney heights of up to 90 feet are possible with only standard chimney supports and no special foundations. Elbow sections, or offsets, are available in 15 degree and 30 degree pieces. Fortunately, they require no trimming or cutting. One of the worst things about building an offset masonry chimney is the need for accurate cutting of fireclay tiles. The larger twist-lock sections (8- and 9-inch diameter) can be adapted for use on 30- and 34-inch heat-circulating fireplaces by adding a masonry-to-metal base plate adapter after the fireplace surround has been nearly completed. The savings in effort and money will make this adaption well worthwhile.

Most installed freestanding fireplaces and wood stoves have about 6 feet of standard smokepipe leading from them. Usually the smokepipe fits into a firestop spacer above and is connected there to an insulated prefabricated chimney (Fig. 13-14). Elbows can be used to bring the smokepipe through the wall at the side of the house instead of through the roof—as long as there is an 18-inch clearance between the smokepipe and any combustible surface (usually the room's ceiling). If a fireplace or stove has been designed for a top-mounted smokepipe (vent), like most freestanding fireplaces and stoves, then the first stovepipe section installed should be the piece which has the damper in it. If a stove or fireplace is designed for a side- or rear-mounted vent, then the first stovepipe section installed should be an elbow, which connects with the straight piece of smokepipe containing the damper. Side- and rear-venting wood stoves without internal baffling can be slightly more efficient than top-venting wood stoves without internal baffling. Less heat is lost straight up the chimney with rear and side vents.

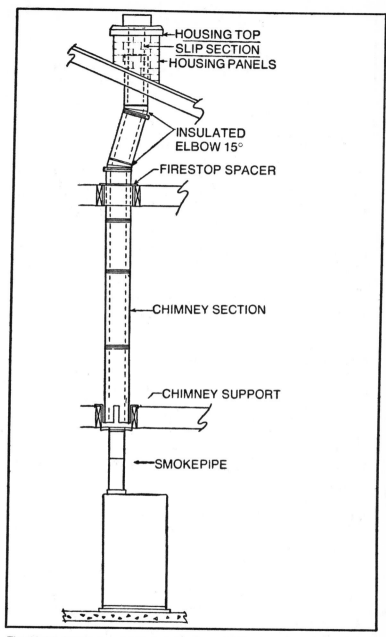

Fig. 13-14. A typical prefabricated chimney installation (courtesy of Heatilator Fireplace).

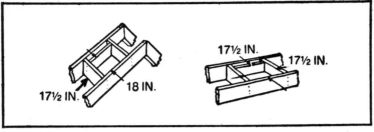

Fig. 13-15. Ceiling and roof openings for Majestitherm.

All the prefabricated chimneys I've seen require only 2-inch clearance from combustible surfaces. They also require firestop spacers every time they pass through a wall, ceiling or roof. If a prefabricated chimney is extra long, or has several flues, it will need chimney supports. You can buy chimney supports from the manufacturer of your particular chimney or you can easily make them of scrap steel at home. Some building codes may require such supports more often than the manufacturer feels is essential, so a quick check with the local building inspector may prevent some later headaches.

All the basics for chimney installation hold true for prefabricated chimneys. You still need the 2 foot clearance above the highest point on the roof within 10 feet of the chimney. The chimney top still must be 3 feet above a flat roof surface. The minimum chimney height (from throat to chimney top) for most prefabricated chimneys is about 15 feet. The chimney cross-sectional flue area must still be from one-tenth to one-twelfth the area of the fireplace opening.

Let's go step by step through a fairly complex prefabricated chimney installation: the installation of the Majestic Majectitherm chimney.

Install the fireplace on the base framing—this will be necessary if there is a raised hearth since it changes the number of chimney sections needed. The chimney will be much easier to position if the fireplace is in its final location.

Use either a plumb bob or a straightedge taped to a level to mark the ceiling over the top of the fireplace. You should take the marks from the outside edge of the chimney installation plate on the firebox. Then mark the outline of the firestop spacer on the ceiling. After that, open up the ceiling using a drill and a keyhole or saber saw. If there is a second floor, repeat the process for the second ceiling. Each opening should be properly framed. The ceiling opening(s) should be 17½ × 7½ inches. The roof opening should be about 17½ × 18 inches (Fig. 13-15).

Once the framing is completed, install the starter section of the prefabricated chimney on the fireplace (Fig. 13-16). On the Majestitherm, the starter section is a triple-walled unit. The three sections (inner, middle, and outer) may be installed one at a time, starting with the inner one, or it may be necessary to hold the

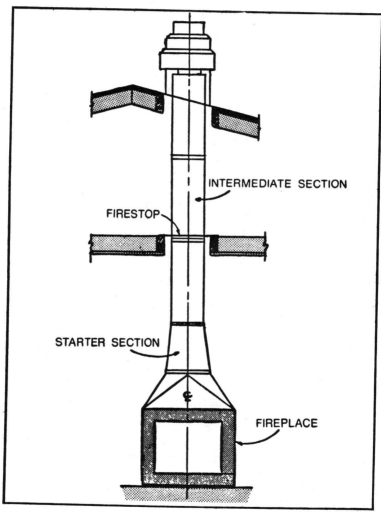

INTERMEDIATE SECTION

FIRESTOP

STARTER SECTION

FIREPLACE

Fig. 13-16. A prefabricated chimney showing the position of the starter section (courtesy of Majestic).

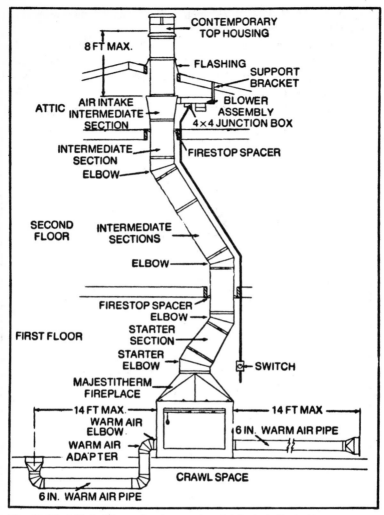

Fig. 13-17. A Majestitherm fireplace and prefabricated offset chimney.

middle and outer sections up toward the ceiling opening while installing the inner section. Make sure that all three sections are firmly engaged (they use snap-lock fasteners) to the proper rims on the fireplace flue outlet.

At each ceiling level install a firestop spacer. Nail it to the joists and headers. No firestop is required at roof level with the Majestitherm. Most other brands will accept some sort of firestop if you wish to install one.

Install the triple-walled intermediate sections in the same manner as the starter section and continue right on up into the attic area (Fig. 13-17). In the attic, install the air intake section. The higher you locate the air intake section, the more heat you will get because heat is gained from the chimney surface. Other manufacturers do not use this sort of triple-wall chimney, nor do they use an air intake section, so the installations are correspondingly simpler.

Whether or not an electric blower is added to the air intake section, the air intake section must have a minimum of 18 inches clearance (height). The chimney must remain perfectly vertical, always maintaining its 2-inch clearance from any combustible surfaces. Attic ventilation should meet minimum FHA standards for proper, safe operation.

If an attic has one-quarter square inch of ventilator for each square foot of attic space, it is properly ventilated. If not, you can install extra vents in end gables or under the eaves. In drastic cases, you can also install a vent in the side or end of the chimney housing.

For the Majestitherm, you must install the air intake section within 8 feet of the top of the rain cap for efficient and proper operation. The closer to the top, the better. In fact, you can install the section in the housing just a couple of feet under the rain cap. The air intake section has a large bell shape on one end. The bell shape must always be at the top of the installation (Fig. 13-18).

In all installation work, make sure that the snap locks are firmly engaged. This holds true for all prefabricated chimneys, whether they use snap locks or twist locks.

Next, install the electric blower. There are predrilled holes in the air intake section. The blower attaches directly these, motor

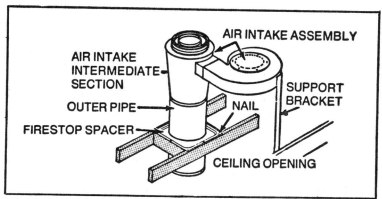

Fig. 13-18. The Majestic air intake assembly.

side down. An adjustable support bracket secures the blower to the floor or to a rafter. Check the blower wheel to make sure it is free turning: simply give the shaft a quick spin to make sure it moves freely and causes no scraping noises.

Install the blower motor according to local electrical codes, making sure that all wiring runs on the *outside* of the firestop spacers. A solid-state control switch and plate are supplied for installation near the fireplaces.

Continue the intermediate sections of the chimney on up until they extend through the roof, at least 3 inches beyond the highest part of your already cut roof opening.

You can select top housings to fit the roof pitch. You can, of course, build your own top housing and chase. If you do, and are using a Majestitherm unit, you'll need to install two louvered vents to allow enough air to circulate down the chimney.

And that's about it. The fireplace and chimney, once they're checked for smoke leaks, are ready to go.

In some installations, whether masonry or prefabricated, one other thing may become necessary: a cricket. A cricket is a section that fits between the roof and the chimney to prevent debris and snow buildup (Fig. 13-19). It is a little peak made to match the pitch of the roof and the wall of the fireplace housing. It is usually, if small, made of 2 × 4 framing nailed together with 16-penny nails. It is covered with sheathing board or exterior grade plywood in the proper weight (three-eighths of an inch for spans of 2 feet or less, one-half of an inch for longer spans). It is then nailed to the roof, covered with roofing paper, then flashed and shingled. In general, crickets are not required if the side of the fireplace facing the roof peak is under 18 inches wide.

CHIMNEY CARE

Chimneys, if properly installed, will cause you no problems for years and years, especially if you clean them once every two or three years. If you use a lot of resinous firewood, you can use table salt to cut down on the amount of buildup. Get the fire roaring hot then toss on a handful. You can also use a solution of table salt and water to wash down the chimney walls every few years, though it's seldom necessary.

A commerical chimney cleaner can clean your chimney better and faster than you can. But it's a lot cheaper if you do the job yourself, and you can easily contain much of the mess.

First, seal off the fireplace opening. If you have an asbestos board cover, simply tape it over the opening. If you use a sliding

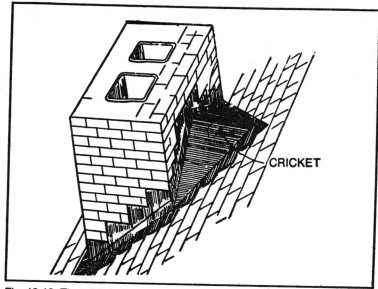
Fig. 13-19. The cricket of a chimney.

glass fire screen, use masking tape to cover the seams and any holes. First, pull the cotter pins holding the damper unit and lift it out onto a pile of spread newspapers.

Next, fill a heavy burlap bag with straw, wood shavings, or any similar substance; the rougher the material of the bag, the better. Tie a brick to the bottom of the bag. From the top of the chimney, lower the bag slowly down the flue, giving the rope a twist every couple of feet. Repeat several times. In a pinch, the burlap bag can often be replaced by a small evergreen tree just big enough to fit snugly in the flue. Be sure to tie a brick to the bottom of the tree.

Be careful climbing back down the ladder with the bag, for the bag will be about as filthy as anything you'll ever see. Bang it around as little as possible, and don't just toss if off the roof unless you enjoy clouds of soot all over your garden. Empty the bag carefully. Then set the emptied bag aside for washing. Burlap bags, though not fantastically expensive, are getting hard to come by in many areas, particularly since many feeds are now packed in plastic sacks. The whole idea of doing this somewhat nasty job yourself is to save money.

If the resin buildup in the chimney is severe, tie a heavy chain on a rope and lower it down the chimney, twirling it hard as it goes down. This procedure will knock a lot of the resin loose, but use it only when the buildup is heavy—it's bound to be hard on mortar joints and fireclay flue tiles.

By the time you store the bag and put the ladder away, most of the mess trapped inside the fireplace will have settled. Sit down and rest for a half hour or so any way. The extra time may not be needed, but if you uncover the fireplace opening too soon you'll have a real dirty house on your hands.

Peel off the tape and remove the fireplace covering slowly and stand it on newspaper. Set everything well to one side and immediately install and close the damper. Any sort of downdraft at this time could be a disaster.

Don't sweep the soot from the fireplace. Vacuum it. If at all possible, don't use your household vacuum cleaner. A shop vacuum cleaner that can be used wet or dry is a much better tool for this job because it allows you to later take the garden hose to it to get rid of some of the internal mess. It also doesn't have an air-permeable bag inside to allow soot to blow all over the house should anything go wrong. In any case, make sure all the vacuum's filters are clean.

Though resinous buildup cannot be *totally* removed by anything other than chipping with a putty knife, the chain treatment should be a big help. Chemical removers, as far as I can discover, don't work very well. Some may combine with the soot to cause a dangerous explosion or a bad chimney fire. The creosote formed by burning wood resins is much harder to remove than soot—a good reason for avoiding resinous wood whenever possible.

Chipping is most difficult in masonry chimneys; they have more rough edges to catch the buildup and are more easily damaged because of mortar joints. Obviously, the interior of any chimney over a few feet high is going to be very difficult to chip free of anything, so your best bet is to hope the buildup remains minor.

In a pinch, you can take down a prefabricated chimney for cleaning, a distinct advantage for those areas where burning wood for fuel means burning resinous wood. You can also take down smokepipes for cleaning. Where sections are only a few feet long, cleaning is much easier.

AN ALL-MASONRY FIREPLACE

I've already covered the complexities of the all-masonry chimney for wood stoves and fireplaces. Now it's time to take a solid look at what is required to install an all-masonry fireplace. If you're a true traditionalist, this is the only kind of fireplace for you. Never mind

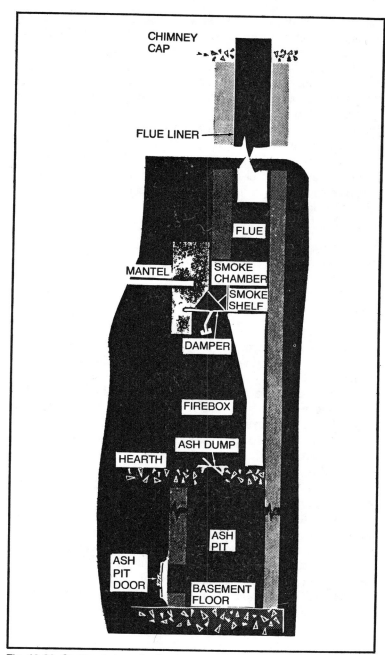

CHIMNEY CAP

FLUE LINER

FLUE

MANTEL

SMOKE CHAMBER

SMOKE SHELF

DAMPER

FIREBOX

ASH DUMP

HEARTH

ASH PIT

ASH PIT DOOR

BASEMENT FLOOR

Fig. 13-20. Sectional view of a fireplace (courtesy of Majestic).

the fact that is usually costs twice as much as any other kind of fireplace. Never mind the fact that it's more difficult to build. *This* is a fireplace!

One of the basic advantages of the masonry fireplace is that you can build it in virtually *any* shape or size. However, you must maintain the proper relationship between flue size, fireplace opening, fireplace depth, and fireplace taper. You must also insure proper throat and smokeshelf sizes and the proper incline for the upper part of the back wall. There are, as you can see, more than just one or two complexities in the construction of the masonry fireplace. Building an all-masonry fireplace is no cinch. There are several components, each one requiring precise planning and careful work (Fig. 13-20).

Footings

The masonry fireplace must have the same kind of footing as that required for masonry chimneys and exterior-wall installations. The footing prevents the masonry from settling improperly or just plain toppling over. Pour the concrete at least 1 foot wide and 1 foot below the frost line.

Hearth

The fireplace hearth should project as least 24 inches in front of the fireplace opening (the floor inside the firebox itself is also called the hearth). The hearth should also project from 6 to 12 inches, at a minimum, to each side of the fireplace. There are so many materials and so many methods of hearth construction that it would be impossible to cover even a majority of them here. Suffice it to say that a hearth should have at least a 4 inches thick base of poured concrete, though a brick hearth can be thinner. Terra-cotta, stone, ceramic tiles and copper tiles can be used as a finish covering, as can any noncombustible material which seems attractive to you and your family.

A hearth flush with the floor is easy to clean out. It also offers a quick and easy sweepup of wood chips and such since they can just be swept directly into the fireplace. The flush hearth is usually easier to install than a raised hearth in an all-masonry fireplace because there's no need to cantilever the poured concrete. The raised hearth offers a warm seat and a convenient spot under which to store a fair amount of wood.

Ashpit

Built under the rear of the hearth, the ashpit offers an easy way to clean out the ashes (Fig. 13-20). Today's ashpits are covered

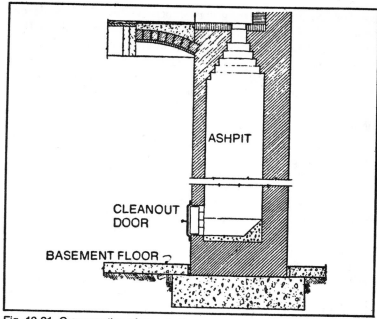

Fig. 13-21. Cross section of an ashpit.

with tight-fitting cast-iron doors which lead into pits made of either ceramic or metal. Cleanout doors are located either in a basement or on an outside wall of the chimney structure. The doors are handy but not essential (Fig. 13-21).

Walls

Most building codes today require that the back and sides of any masonry fireplace be made of solid masonry or concrete at least 8 inches thick, lined with firebrick or other approved material. Approved materials include steel linings, which are supplied by heat-circulating firebox manufacturers.

Some codes allow you to eliminate linings of firebrick or other such materials if you use 12-inch solid masonry (or reinforced concrete) walls. Regardless of codes, I would not recommend eliminating the lining because heat from a wood fire can have a disastrous effect on any masonry wall, causing the fireplace to need almost constant repair.

Jambs

Fireplace jambs are those upright parts on the sides of the fireplace opening. You should design the jambs to provide construction stability and a pleasing appearance. With a 3-foot wide or smaller fireplace opening, I'd recommend a 12-inch width for the jambs. Proportionate increases in that width should be made as the fireplace openings grows in size. The jambs can be faced with ornamental tile or brick, but wood should be avoided at any point closer than 6 inches to the fireplace opening.

Lintel

A lintel is the support for the masonry which crosses the top of the fireplace opening. With an opening of 48 inches or less you can use one-half of an inch flat steel bars for the lintel. You can also use $3\frac{1}{2} \times 3\frac{1}{2}$ by one-quarter inch angle iron. Of course, you can use specially designed damper frames too. Heavier lintels are needed for wider openings. If a masonry arch is used, the lintel must be increased in size by a large enough amount to resist the thrust of the arch. It might be a good idea to check with an architect, architectural engineer, or civil engineer to find out what exact size you need to prevent collapse. Any fee should be minimal since it will only take a few minutes work for the professional to come up with the figures you need.

Throat

Any improperly shaped throat will totally ruin the draw of the fireplace. The sides of the fireplace should rise vertically to the throat, which must begin from 6 to 8 inches above the lintel bottom. The cross-sectional area of the throat must at least equal that of the flue, and the length must at least equal the width of the fireplace opening. Five inches above the throat, the sidewalls start sloping in to meet the flue. A large fireplace must have a much greater slope than a small fireplace, and the slope is absolutely determined by the throat size.

Damper

Dampers are cast iron frames with a hinged lid that can be opened or closed to vary the size of the throat opening (Fig. 13-22). In older fireplaces dampers were not only often installed incorrectly, but were often not installed at all. It's usually the undampered fireplace which operates at a net heat loss for the house, unless a fire is kept roaring all the time. The damper not only controls the draft for the fire in the fireplace but helps to keep already heated air from flowing up the chimney when the fireplace is not in use. For that reason, a tight-fitting damper is essential. You can also adjust the throat opening to reduce heat in the room while the fire is going. Though a roaring fire with resinous wood may need a full throat opening, a slow-burning hardwood log can often last all night if the throat opening is cut down to an inch or so. A good damper can keep the bugs out of the house in the summer. If you think that's a

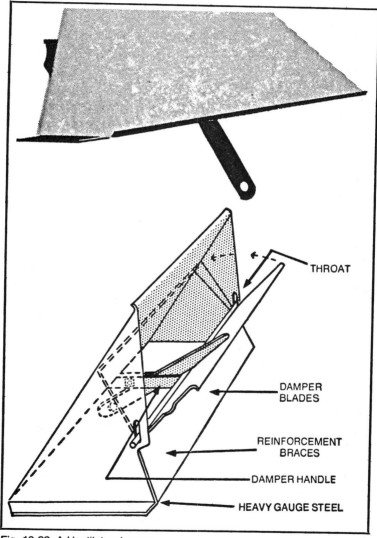

Fig. 13-22. A Heatilator damper.

facetious remark, you've never liven in an insect-ridden area with three fireplaces in the house and not a damper among them. I have! Some of today's dampers are also designed to take the place of lintels and can save a fair amount of worry in fireplace building.

Smoke Shelf and Smoke Chamber

The smoke shelf has one simple and important reason for existing. It is there to prevent downdrafts (Fig. 13-23). You make a smoke shelf by moving the brickwork at the top of the throat back to the line of the flue wall for the full length of the throat. Its depth from 6 to 12 inches, depends on the depth of the fireplace. The smoke chamber is the area from the top of the throat to the bottom of the flue. Smoke shelves and smoke chambers should be plastered with at least a one-half inch cement mortar.

Flue

Properly sizing the area of the fireplace opening to the flue, taking into consideration the height of the flue, is the single most important factor in preventing smoking fireplaces. Remember, if a flue is 15 feet tall or more, its cross-sectional area should be about one-tenth the area of the fireplace opening. If a flue is unlined or if it is less than 15 feet tall, the flue must increase in size to one-eighth of the area of the fireplace opening. We've pretty much covered the rest of the information you need for flue selection in the chapter on chimneys. But as a reminder, stay as close to the above relative sizes as possible because too large a flue area can cause nearly as many problems as too small a flue area. Stay with the standard size rectangular fire clay flue liner whenever possible because the rectangular flue is more efficient than the round.

And that's what a fireplace is all about. The parts involved are common, in one way or another, to all fireplaces, but it's only with the all-masonry fireplace that you need to get involved with shaping and building your own. After you've decided on placement, poured the footings, and gathered all the other materials around you, the real work begins.

Brickwork

Making an accurate estimate of the number of bricks you'll need for your masonry work can save you a lot of money. Bricks are based on 3¾ inch modules. However, this sizing is not exact. You'll certainly find it pays to check the size of bricks you buy from local

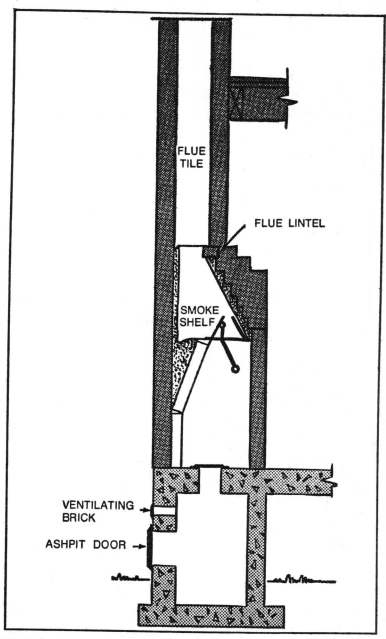

Fig. 13-23. A cross section showing the position of the smoke shelf (courtesy of Majestic).

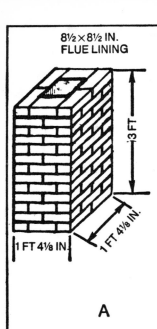

8½ × 8½ IN. FLUE LINING

3 FT

1 FT 4⅛ IN.

1 FT 4⅛ IN.

HEIGHT OF CHIMNEY	NUMBER OF BRICKS
1 FT 0 IN.	30
2 FT 0 IN.	54
3 FT 0 IN.	78
6 FT 0 IN.	156
9 FT 0 IN.	244
12 FT 0 IN.	312
15 FT 0 IN.	390
18 FT 0 IN.	468
21 FT 0 IN.	546
24 FT 0 IN.	624
27 FT 0 IN.	702
30 FT 0 IN.	780

A

1 EACH 8½ × 8½ AND 8½ × 13 IN. FLUE LINING

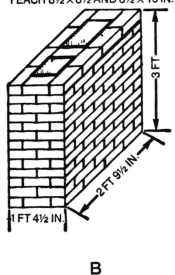

3 FT

2 FT 9½ IN.

1 FT 4½ IN.

HEIGHT OF CHIMNEY	NUMBER OF BRICKS
1 FT 0 IN.	55
2 FT 0 IN.	99
3 FT 0 IN.	143
6 FT 0 IN.	286
9 FT 0 IN.	429
12 FT 0 IN	572
15 FT 0 IN.	715
18 FT 0 IN.	858
21 FT 0 IN	1001
24 FT 0 IN.	1144
27 FT 0 IN.	1287
30 FT 0 IN.	1430

B

Fig. 13-24. Brick quantity for four masonry chimneys based on standard brick size: 2¼ × 3¾ × 8 inches, with one-half inch joints (courtesy Heatilator Fireplace).

TWO 8½ × 8½ IN. FLUE LININGS

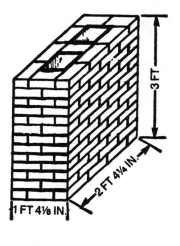

C

HEIGHT OF CHIMNEY	NUMBER OF BRICKS
1 FT 0 IN.	50
2 FT 0 IN.	80
3 FT 0 IN.	130
.6 FT 0 IN.	260
9 FT 0 IN.	390
12 FT 0 IN.	520
15 FT 0 IN.	650
18 FR 0 IN.	780
21 FT 0 IN.	910
24 FT 0 IN.	1040
27 FT 0 IN.	1170
30 FT 0 IN.	1300

1 EACH 8½ × 13 IN. AND 13 × 13 IN. FLUE LINING

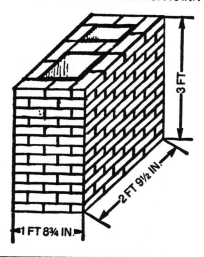

HEIGHT OF CHIMNEY	NUMBER OF BRICKS
1 FT 0 IN.	63
FT 0 IN.	113
3 FT 0 IN.	163
6 FT 0 IN.	326
9 FT 0 IN.	489
12 FT 0 IN.	652
15 FR 0 IN	815
18 FR 0 IN	978
21 FR 0 IN.	1141
24 FR 0 IN	1304
27 FT 0 IN.	1467
30 FT 0 IN	1630

suppliers to make sure they're not off an eighth of an inch or so. If you do find that your brickmaker produces bricks one-eighth of an inch smaller than the norm, then you'll need approximately 3 percent more bricks. If the bricks are one-eighth of an inch larger, you'll need, of course, 3 percent fewer (Fig. 13-24). You should allow one-half of an inch for each mortar joint.

Use the mortar mix recommended by your local brickmaker and then work to get the right mix feeling. Plasticity of the mix is important for easy handling, but the dry mix must be just right to provide strength. Generally, with masonry cement, you can add 4 cubic feet of sand (in damp, loose condition) to one-third of the sack of cement. Mix it until the contents are evenly stirred together. Add the water until your mixing hoe can be shaken clear of the mortar. Add the water as slowly as possible; that is, don't get in a rush and dump in a lot of water because the less water you use, the greater the strength of the dried mortar.

If the brick is not wet enough, it may draw off too much water from the mortar. If this happens, the mortar won't cure properly and you'll end up with weakened joints. Check wetness with a sprinkle test. Sprinkle a few drops of water on the brick and look at your watch. If the water is absorbed in a minute or less, the brick needs to be wet down. Use a hose to wet down the entire brick pile, keeping the water flowing until you get a good runoff from all sides of the pile. Wait a few minutes for the surface water to soak in or evaporate before using the brick.

If you were a professional bricklayer, you'd be able to pick up enough mortar in a single trowel load to place from three to five bricks. If you've never worked with brick before, your best bet is to pick up a lot less mortar, trying to start with no more than two bricks per trowel load. Toss the mortar onto the bricks with a flick of the forearm muscles as the trowel is turning through a half circle. End the flick with the trowel upside down. Use the trowel point to spread the mortar into a layer that's a bit thicker than one-half on an inch. Scrape off any overhang on the sides with the trowel edge. Each brick, as it is laid, must have its end covered with mortar (called buttering) so that the vertical joints will also be filled with mortar.

Butter each brick and then lay it to the right of its final position (if you're working from left to right). Use a downward, slanting motion to slide the brick into its final position. Tap the brick with the trowel until it aligns with the previously laid bricks. As each brick is so laid, you'll have to scrape off the mortar which is pushed out by the laying operation.

To keep your lines straight, use a line for each row of brick. You can tie the line around the already laid corner bricks. Brickwork always starts at the corners, which are built from five to eight rows high before any interior bricking is done. The line should fall about one-thirty-second of an inch outside the brick wall and should be drawn as tight as possible. If the mortar in the corners has not set, be careful not to yank the corner brick out of the wall. No brick should touch the line. Mason's cord and mason's distance pieces are available at most building supply and hardware stores and are very useful. The novice bricklayer should check the level of each course of bricks several times, possibly as often as every three bricks. Though constant checking is a time-consuming chore, there's nothing worse than having to tear a lopsided wall down and start over again.

For fireplaces and chimneys, you should lay the first course of bricks on the footing or on the block wall on top of the footing if the footing is so deep the first course of bricks won't reach ground or hearth level. All such foundation work should be done with blocks or poured concrete because it is a lot cheaper than laying bricks below grade. Lay the first row of bricks dry to figure out how many bricks will be needed and where to make any cuts that might be necessary. Of course, you should allow for a one-half inch mortar joint between bricks so that any cuts will be accurate. A few chunks of folded and taped cardboard, premeasured to one-half of an inch can help maintain this distance as you move along the wall.

After you get your measurements, lay that first course again, this time with mortar. Build up the corners five or six rows high and begin laying the courses between the built-up corners. When you reach the top of the corner, repeat this process until you reach your final height or the point where the brickwork must change direction or shape.

Just about anyone can lay those fine smooth joints you see on professional brickwork. A concave pointing tool is used to press squeezed out mortar back into the joint and to scrape off any excess. The tool can have a V-shape or a U-shape. The compressed mortar in the joint aids waterproofing as well as appearance.

A masonry firebox will need three-eighth inch fireclay joints. Even cement masonry won't withstand the heat of the firebox for very long, so the fireclay is essential to a good long-lasting job. Firebrick is usually 9 × 4½ × 2½ inches. You can lay it with either the 2½ inch or the 4½ inch face exposed. Local building codes should be checked first to make sure the firebox wall will be thick enough if the 4½ inch face is exposed. At any rate, the lower back and the sides of the fireplace will present few bricklaying problems. When you reach the proper height for the fireplace back to start its forward slant, care becomes very necessary. You'll probably have to slow down to give the fireclay joints more time to start to set or the whole works may tumble out on you. An experienced mason can do this sort of work in one shot, and getting it right demands a bit of experience. Patience is a good replacement for novices.

If you expect too much trouble, you can build up a form from plywood; you can then lay the bricks from outside the firebox. The form is just two pieces of plywood, with the top one angled properly for the slant of the back wall. The form should be braced with 2 × 2s. This causes some problems with getting smooth mortar joints on the fire side of the firebox, but this can be partly helped by using less mortar on the front side of the brick so that no extra is squeezed out.

After you build up the firebox the proper height, imbed the damper and install the lintel (if needed) across the fireplace front. Dampers are available from many manufacturers in many sizes, but they all have one thing in common: they need to be set in a bed of rock wool insulation to allow the metal of the damper to expand and contract. Your damper manufacturer will specify a flue size to fit the damper you purchase.

All firebox joints should be trowel finished. That is, they should be scraped off flush with the face of the brick using the side of the trowel.

Corbeling is the method used to slant in, or slant out, the brick walls surrounding a fireplace or chimney. For slanting *toward* the fireplace, this is done by slipping the top row of bricks 1 inch off the flush line towards the fireplace then repeating the process until you reach the point where you wish to continue upward in a straight line again (Fig. 13-25). If a great deal of weight must be supported, you should consider doing the corbeling with a second row of bricks *inside* the offset.

As the fireplace grows, the spaces between the outer brick walls and the firebrick walls should be filled with masonry rubble, such as broken bricks, dried cement and small rocks.

After you complete the fireplace, it will require at least one week to cure, though even then the first few fires must not be roaring, leaping demons of heat. Small fires can help the curing process of the mortar and the fireclay, but roaring blazes can upset the process and create smoke leaks.

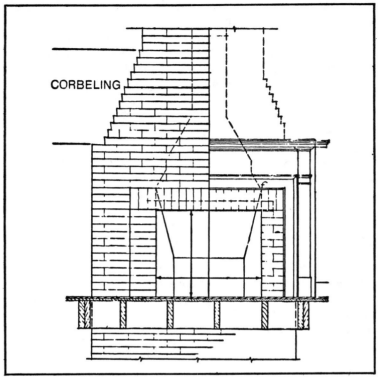

CORBELING

Fig. 13-25. Corbeling on a brick chimney.

ROCK MASONRY

From prefabricated fireplace systems to brick (or cement block) masonry to rock masonry—that's the order of difficulty in building fireplaces. For best results you must build the rock masonry fireplace around a firebreak hearth and firebox with a fireclay flue. But it is possible to save a fair amount of money by working with native stone found on or around your own property. Brick prices vary around the country, but one thing can be said at all times: bricks are expensive. Rocks are everywhere and can make attractive and sturdy fireplaces and chimneys.

There are, unfortunately, a few catches which make rock masonry more difficult than brick masonry. First comes the problem of estimating the number of rocks you need. It's close to impossible using found field stones to make an accurate estimate. Brick estimates are simple, for each brick is of a uniform size and the mortar joints are also uniform. Rocks close to 1 cubic foot are on the not-recommended list because a lot of rocks weigh from 150 to 180 pounds per cubic foot, which is kind of heavy to work with, even with scaffolding. Bricks may have to be trimmed for fit, but unlike rocks they don't have to be chipped to fit a mortar joint. Rocks have to be trimmed at certain fault areas and then slowly chipped to size. This sort of work requires a devilish amount of splitting and chipping. Still, it is possible, and if you've got more time than money, it can be much more economical than many other installations.

Rock masonry fireplaces require the same kind of footing as any other masonry fireplace.

Two types of rocks are available; the igneous rocks, granite and basalt, which are more difficult by far to work with but generally are a lot stronger; and the stratified rocks, limestones and shales, which are much easier to work with but are susceptible to moisture deterioration. The colder and wetter your climate, the more likely you are to have problems with this sort of deterioration. The igneous rocks are usually available in machine-cut styles, but this again runs your expenses right back up there with brick.

Ease of working depends on the weight of the rock. Granite is a very heavy stone (about 170 pounds per cubic foot) and is matched in weight by limestone. Sandstone is also heavy but not as heavy as granite: Sandstone weighs about 150 pounds per cubic foot. Slate is heaviest of all at around 175 pounds per cubic foot.

Rock is laid pretty much the same way brick is laid. However, with rock masonry there's no need to level each course. Rocks are usually easier to lay if you trim the face side flat so you can use a line to keep the stonework as close to vertical along its face as possible.

You should give rocks the same sprinkle test used on bricks. If a few drops of water disappear into the rock in a minute or less, soak down the entire pile.

You'll need at least twice as much mortar as you would need for a comparable brick wall—probably a great deal more, depending in large part on how tight you plan to make the mortar joints and how closely you trim the rocks. You can use small rocks to space out the joints to cut down on the amount of mortar you use. Mortar ingredients for rock masonry are the same as those for brick masonry. Random-rubble masonry is the most difficult rock masonry to lay, but trimmed stratified-rock masonry is the easiest because it closely resembles the course-by-course laying of bricks and concrete blocks. Random-rubble masonry is done with junk rock or brick—bits and pieces of rock or brick laid in a random pattern.

The mason's, or bricklayer's, chisel is an absolutely essential tool for rock masonry work. You'll probably also want a maul or small sledge. For most people, a 2 pound maul is adequate, though exceptionally strong people may prefer a 3 pounder. Look for cleavage lines (even in granite) because rock is more easily broken and trimmed along these natural lines. Use a mason's, or bricklayer's, hammer for the fine trim chipping of stone.

In the actual construction of a wall, use longer stretcher stones about every 6 square feet to make sure that wall strength is maintained and that all parts of the wall are tied together. You can align the wall precisely by using a stake at each end of it. Stonework doesn't allow for wrapping the line around a corner as you do with brick work. Stretch a line tauntly from stake to stake at the proper height and as close to the front surface of the stone as you can get without touching the stone. Of course, you must lift and level the line as the wall moves up. If you must lift a stone out and reset it, clean the mortar from both it and any stones where it had been set. Use fresh mortar for the resetting. Wash all stones before using because dust will prevent the formation of a proper bond between mortar and rock.

Select the flattest rocks for the final course on a chimney and any final courses at other spots. Then use sufficient mortar to get a level top on the construction. Use a jointing tool to tool the mortar joints; clean away any mortar which drops on the stones. As a substitute for a jointing tool, use a stick with a smoothed end to tool mortar joints. Use a rough, wet sponge to clean off any mortar on the stone facing; dry mortar is very hard to remove and almost always leaves a mark on stone.

SOLVING FIREPLACE PROBLEMS

There are not all that many things that can make a properly constructed fireplace start pouring smoke into a room, but there are many mistakes made in the design and construction of fireplaces which can cause them to smoke from the day a fire is lit in them. There are cures—some mild, some strong—for almost any of these problems.

I'll first discuss leading flue/chimney problems. The introduction of smoke into a house can be very dangerous, so you should run a smoke leak test occasionally on just about any fireplace or wood stove that gets extensive use. A leak test every three or four years should be sufficient. After you build the smudge fire, go over the entire length of the chimney looking for leaks. The dense smoke should be immediately visible in the case of most leaks, but you may have to use your nose to track down tiny leaks. Small leaks over a long period of time can become nearly as dangerous as larger leaks. Tar paper in the smudge fire is a big help when you have to sniff for leaks, though it does tend to smell up the house a bit if the leak is not found quickly. If the area of the leak is easily accessible, simply apply cement mortar to the joint until the leak is sealed. If the fire goes out while the mortar is being mixed and applied, a recheck is a good idea. Some leaks may not only be hard to locate, they may also be difficult to get to for repair.

Leaks can be behind walls, inside closets and so on. For this reason, you should use great care when building any chimney. Having to tear out walls to repair a one-sixteenth of an inch leak in a mortar joint can spoil anyone's day. If the leak cannot be repaired, you must *not* use the fireplace or stove.

Fireplace smoking into a room is less dangerous than chimney leaks, but it can be among the most irritating things to ever happen to anyone. You're sitting there, just toasting your feet with the blaze, sipping a glass of brandy, and great rolls of smoke pour into your face. Start your troubleshooting on the outside. Get on the roof and make sure that the chimney top is at least 2 feet above any obstruction within 10 feet of the termination (on a peaked roof). Make sure the chimney of a flat roof is 3 feet high.

If chimney height is not the problem, look down the chimney with a flashlight (use a powerful one or the light won't reach very far) to check for internal obstructions that might have dropped into the chimney.

Downdrafts

If this is not the case, take a look at the trees surrounding your house. Trees can cause wind currents to vary in such a way that you'll get an occasional downdrafts that can pour right down your chimney. Trees that were far enough away from the house when the fireplace was built may take only a few years to send out branches which cause problems. A chain saw and some rope is the answer here, but if you're not an experienced chain saw user please don't go cutting while you're up in the tree. Hire a professional. Of course, you can cut the whole tree down if necessary. You can also build a chimney cap to keep those downdrafts out of your chimney. As mentioned in the chimney section, extending the flue lining 4 inches beyond the top of the chimney and then cementing a rain slant or bevel to within an inch or two of the flue's top will also help to prevent downdrafts by cleaning up the air eddies around the chimney top. This won't work too well, though, when a bluff or hill too close to the house causes the downdrafts. In such cases a chimney cap is a better solution.

If downdrafts are caused by land conditions, a chimney several feet higher with a cap or hood may be the answer. Nearby high buildings can also cause downdraft difficulties. In the case of both high hills and high buildings, sometimes the only solution is to build the chimney on up until it's out of the turbulence which causes the downdraft.

Improperly shaped parts of a fireplace or chimney cause as many, or more, downdraft difficulties as external air current twists and turns. In some cases, some sort of reconstruction of either the chimney or the fireplace is the only cure. If the flue opening is too small for the fireplace opening area, there is a reasonably simple cure. Install a fireplace hood that drops down over the front of the fireplace opening, cutting the area by enough to fit the flue size. If you don't know the flue size you'll either have to climb a ladder or slither into the fireplace to take a measurement. Incorrect flue sizes can exist even on a factory-built fireplace but not on zero-clearance styles because the manufacturer and distributor sell them as packages with the proper flue sizes.

Chimney Hood

Fortunately, a hood can be made of many decorative metals such as copper, aluminum, or sheet steel, in any shape which corresponds to your home decor. So find out the flue size; measure the fireplace opening. Then build a fireplace hood which shuts out the needed areas from the fireplace opening. A narrow metal strip beneath the breast of the fireplace (where the lintel is installed) can do a good job. You can lay a second course of firebrick with fireclay mortar on the firebox to remove a good portion of the fireplace opening area. New firebrick can even be laid up the sides of the fireplace to cut down on the size. If you need a really drastic reduction in size, the hood, the refractory hearth, and the new sides can all be combined, but it's probably best to start out with only the hearth or the hood and see if that works before doing any work which may be unnecessary.

Ventilation

Too tight a house can also cause a fireplace to smoke. If you have a series of appliances in a tightly caulked, insulated house, the added need for combustion air when a fireplace is used can draw air down the flue and cause smoke to billow out into the room. In fact, this sort of tightness can cause problems even without the fireplace being lit. In a tightly sealed house, the chimney is forced to feed outside air to the stove, fireplace, hot water heater, or any other appliances using air for combustion. A tight-fitting damper will prevent much of this loss of heat. Remember that as outside air is drawn into the house, heated house air will pass on up the chimney, causing a net heat loss for the home. But the only solution for a fireplace smoking from too little air for combustion is to provide more air. This may mean opening the doors to one or two more rooms, or it may require opening a basement window. In some cases, you may find it necessary to install a ventilating grate of the type used for hot air heating systems in the room with the fireplace. The grate can lead either to the basement, another room, or outside. A grate leading to the outside is recommended in only the most extreme cases. The grate should be kept tightly closed when the fireplace is not in use.

In any case, if this lack of air situation occurs, the house is probably *too* tightly sealed. Some ventilation is always needed, but with modern worries over ever rising fuel costs, more and more houses are becoming almost hermetically sealed during the winter months.

A Double Flue

A double flue in a single chimney can cause what seems to be many problems but is often one problem with several causes. In a poor chimney design, one flue can suck smoke from another down into itself—right on out into the room when the fireplace is in use. Or an operating flue can have its smoke forced down the opening of one not in use. Sometimes the only cure for all these ills is to reconstruct the chimney so that the flues are at least 4 inches apart, separated by at least 4 inches of solid masonry, with one flue at least 4 inches above the top of the other. If the trouble persists, it's being aided by local down currents, and the chimney will need a cap over one or both of the flues (Fig. 13-26). Since the flue passing smoke can suddenly, with the correct wind currents, become the receiver of smoke, the capping of all flues is recommended.

Construction faults ranging from sloppy workmanship to poor materials can play a big part in fireplace problems. If the brick is perforated, or too porous, and the workmanship on the mortar joints is a bit sloppy, the possibility of flue to flue leakage is increased. Without tearing out the whole works and starting over,

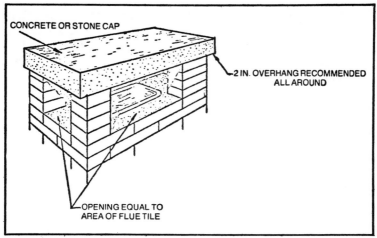

CONCRETE OR STONE CAP

2 IN. OVERHANG RECOMMENDED ALL AROUND

OPENING EQUAL TO AREA OF FLUE TILE

Fig. 13-26. A typical rain cap for masonry chimneys (courtesy Vega Industries Inc.).

there's not much you can do about perforated or porous bricks, but sloppy mortar joints on the exterior of the fireplace and chimney can and should be repaired.

Sloppy mortar joints on the interior of the flue are a bit more of a problem, but they can usually be handled without major reconstruction of the chimney. Again, a burlap sack becomes a handy assistant. Make sure the bag, full of straw or wood chips, is a tight fit. Remove the damper by pulling the cotter pins and tilting it out. Set the burlap sack on the refractory hearth and drop a moderately heavy rope down from the top of the chimney. Lift the sack almost to the flue inside the smoke chamber and cover its top heavily with mortar. You may need a helper at this point. Draw the sack upwards very slowly, twisting the rope as it comes. Repeat as needed until flue-to-flue leakage of smoke is stopped. After each pass you might want to run a clean tight-fitting sack through to keep the mortar from building up on the inside of the flue lining.

Missing Parts

Though this may be a rash step, I'm not going to cover such problem causers as missing smoke chambers, missing smoke shelves, off-center flues, and the like. The sizes and parts of already built fireplaces and stoves are easily checked from information I've already supplied. The only solution in such cases is to tear the unit down and rebuild it correctly. With careful checking, you can often keep from having to tear out too much of the fireplace. In fact, it's often possible to tear out the face of an old fireplace and install a heat-circulating unit. With careful teardown, the old brick will be usable after cleaning and will add a great deal to the appearance of the new fireplace.

A Choked Flue

A mortar-choked flue can cause smoking, but more often the choked off flue will be filled with other kinds of obstructions. You may find a bird's nest. Squirrels and other rodents sometimes build nests in unused flues. If you've built your fireplace the easy way, with a straight flue with no offset, the removal of such obstructions is no great problem. You simply pull the cotter pins holding the damper control in place, lift it out, and use a long stick to poke any obstructions down. They'll fall all the way to the hearth, or lodge on the smoke shelf where they can easily be lifted out. In a chimney with one or more offsets you could run into real difficulty. It's possible to get junk lodged in these offsets that is almost impossible to budge with a long pole of any size. If that happens, try a bundle of logging chains on a rope. And if that doesn't work, you're in trouble. The only solution remaining is to rip out the chimney at that point, remove the obstruction, and rebuild the chimney. Screening off the top of the chimney is a good way to keep most kinds of debris out of the chimney.

Improper smoke chamber construction can also obstruct the flue. The final course of bricks which hold the flue tile should be offset just enough to fit the flue lining without cutting into the opening of the flue. Any bricks which do cut into the flue opening will have to be removed and reset.

A clogged ashpit can also cause problems. Usually this problem will only come about if the wall around the ashpit leaks water. In a correct installation, the ashpit door will be large enough to allow you to take a poker and jam around in the mess to clean it out. If clogging happens more than once, consider doing a waterproofing job on the wall surrounding the ashpit—but only after checking the flashing at the top of the chimney for possible leakage. Defective flashing can allow water to drain right down the sides of the chimney into the ashpit.

Damage from ice, snow, and rain can cause a lot of problems in older fireplaces *and* in some newer ones. Because there is a slightly different contraction rate for each of the various materials used in chimneys, you can expect some cracking of the mortar joints and of the beveled rain cap over a period of years. Improper mortar mixtures, poor construction techniques, and too short a curing time can cause these cracks very quickly. An annual check of the cement rain bevel at the top of the chimney should be routine. This rain bevel is the most likely point for the entry of moisture, which can then freeze and expand the cracks, causing rapid disintegration. Make periodic checks of all mortar joints on the exterior of the chimney.

Use of andirons or a firebasket can help some flue size problems, though you'll usually need to take other corrective measures. Still, either accessory will make fires easier to start and keep them burning more brightly. The firebasket is the better of the two at this job, as is a grate, since both provide more support than do andirons, which often allow the middle of the fire to collapse.

Fire Place Depth

Fireplace depth is of great importance too. Too shallow a fireplace will prevent you from building much of a fire without bringing smoke and flame into the room. Too deep a fireplace will provide its own share of smoke problems because the front of the fireplace may stay considerably cooler than the rear of the fireplace, causing eddies that throw smoke into the room. A shallow fireplace throws more heat into a room than does a deep one with the same fireplace opening area. A good minimum depth is about 16 inches for even the smallest of fireplaces. This depth allows a reasonable size fire to burn without the danger of burning logs rolling out onto the floor. A fair maximum depth, for fireplaces up to 6 feet wide and 40 inches high, is about 26 or 28 inches. As an example, the Majestic heat circulator, 54 inches wide and 31 inches high, offers a 19 inch depth, which the company considers more than enough. Heatilator's 60-inch unit offers a 21¼-inch depth. Since both these units were designed to increase fireplace efficiency by circulating more heat, they were tested to insure the fires burned properly.

A fireplace of more than half the depth of the fireplace opening width can waste heat and has wasted money in building materials. Masonry fireplaces that are too deep can have a row of firebrick added to their backs to take up as much as 4½ inches of the space being wasted. Make sure there is no interference with the flue, though.

Hollow Clay Tile

Although it is a bit harder to locate than brick, hollow clay tile—also often called structural clay tile—can offer some advantages when used in a job that might otherwise call for brick. It can be used as a back up for a brick wall or as the total wall.

Hollow clay tile is made of the same materials as brick, (burned clay or burned shale) but it comes in much larger sizes. Clay tile has central cores that are larger than found even in cored brick. All hollow clay tile is made by the extrusion method. It is pressed through special steel dies and then cut to specific lengths. This provides a quite smooth surface, which can be an advantage with certain projects.

Terminology differs somewhat, as the exterior of the hollow clay tile is called the shell, while the inside solid partitions between the cores is called the web. Shells are at least three-quarters of an inch thick, while the web will be no less than a one-half inch thick.

The greater size of the hollow clay tile allows for more rapid construction of a project than does brick. Load bearing types, of which there are two, can be directly substituted for brick construction if you prefer. Various finishes are available so that you have a choice of finishes for your job.

Grade *LB* hollow clay tile is load-bearing and is used masonry construction which will not be exposed to weather. Examples are interior construction and backing for exterior brick walls. Actually, such a wall could be faced with stucco on the exterior. Plaster, gypsum wallboard, panelling, ceramic tile, or just about any other form of wall finish could be used on the interior. If a stucco or similar exterior finish is used, it must be at least 3 inches thick. A brick or stone facing is usually more practical. Standard metal brick ties are used to join the two portions of the wall, which can be of hollow construction to enhance insulation properties and cut down even further on the transmission of sound.

INTERIOR PARTITIONS

Grade *LB* tile also provides a good material for interior partitions that must bear loads. It will also usually provide better sound and heat insulation than will standard wood framing methods.

If the tile is laid vertically, it can be used as conduits for electrical and plumbing runs and some tile is made with large enough core dimensions to serve as heat ducting. If a plaster wall finish is prefered, the tile comes specially cored so that no lath is needed for the paster to adhere. However, you are likely to find some difficulty in locating anyone who knows how to plaster walls. Therefore, it is often best to use an adhesive to glue gypsum wallboard or wood panelling directly to the tile. If it seems likely that at some later date you will want to hang pictures, shelves or other such items from the wall, I recommend that you frame with 2 × 2s and nail the wallboard or panelling to that. On 2-foot centers, such framing goes up rapidly and adds minimally to the cost. This provides enough depth before the tile is reached to allow for the insertion of various types of wall anchors without having to use a ceramic drill to penetrate the tile. To me, this type of construction seems almost ideal for a basement where one room could be used as a den or bedroom and the other side as a workshop or recreation room, whether or not the face framing of 2 × 2s is used.

Unexposed tile can be bought scored on all sides. Other types have one or more surfaces that are quite smooth, giving the appearance of face brick, in much larger units. Some hollow clay structural tile is also available with specially glazed surfaces, similar to ceramic tiles, so that it is easily cleaned. Many colors are available and these types can be quite attractive, while providing a stain proof surface.

Grade *LBX* is similar to Grade *LB* but is burned to a greater degree so that it can be exposed to weather with no other facing.

Two other classifications apply to hollow clay tile. Type FTX has a surface free of defects (and would be the type to be glazed). This type is quite smooth. Type *FTS* has surface defects that don't detract from strength, but do detract from appearance. This type is most suitable for walls to be faced in some manner.

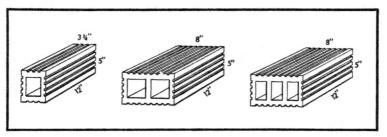

Fig. 14-1. Three standard shape and size structural clay tiles.

SIZES AND SHAPES

As with brick, structural clay tile comes in many sizes and shapes. However, it is primarily based on 4-inch nominal module. When mortar joints are added, the tile will be some multiple of 4 inches.

As with brick, the compressive strength of structural clay tile is much greater than its tensile strength. Because of the variety of sizes of cores and web thicknesses, there is a wider variation in compressive strengths. Tiles having smaller cores and thicker webs are the strongest, assuming the same type of clay is used.

INSULATION PROPERTIES

While fire resistance is somewhat lower than it is with solid brick units, structural clay tile provides greater heat and sound insulation because of the dead air space provided by the cores. Insulating properties, both sound and heat, are not as good as with a hollow brick wall since the webs do serve to transmit both sound and heat from one side of the wall to the other. But used in combination with brick for a hollow wall, structural clay tile provides almost complete sound dampening and excellent heat insulation.

Perhaps the greatest savings to be expected when using hollow structural tile is in the cost of the footings needed. A solid brick wall weighs about four times as much as does a 6-inch wide hollow tile wall and almost three times as much as a foot thick tile wall.

Figures 14-1, 14-2 and 14-3 show some standard shapes of structural clay tiles for both end construction and side construction jobs. Side construction simply means that the tile is designed to be laid on one of its shell faces, while end construction clay tiles are designed to be laid on their web faces to leave vertical cells, or cores, which can be used for plumbing, wiring or other purposes.

Tables 14-1 and 14-2 show the material amounts required for both end and side construction walls, using various size tiles. These

272

Table 14-1. Hollow Clay Tile Wall Construction Material.

Wall thickness	4 inches		6 inches		8 inches		10 inches	
Tile size	4 × 12 × 12		6 × 12 × 12		8 × 12 × 12		10 × 12 × 12	
Wall area, sq. ft.	Number of tile	Cu. ft. mortar	Number of tile	Cu. ft. mortar	Number of tile	Cu. ft. mortar	Number of tile	Cu. ft. mortar
1	.93	.025	.93	.036	.93	.049	.93	.06
10	9.3	.25	9.3	.36	9.3	.49	9.3	.4
100	93	2.5	93	3.6	93	4.9	93	6
200	186	5.0	186	7.2	186	9.8	186	12
300	279	7.5	279	10.8	279	14.7	279	18
400	372	10.0	372	14.4	372	19.6	372	24
500	465	12.5	465	18.0	465	24.5	465	30
600	558	15.0	558	21.6	558	29.4	558	36
700	651	17.5	651	25.2	651	34.3	651	42
800	744	20.0	744	28.8	744	39.2	744	48
900	837	22.5	837	32.4	837	44.1	837	54
1,000	930	25.0	930	36.0	930	49.0	930	60

*Quantities are based on ½-inch-thick mortar joint.

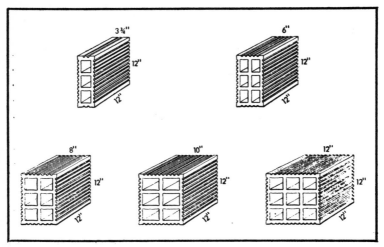

Fig. 14-2. Five end construction style structural clay tiles in standard shapes.

figures are based on the use of one-half inch mortar joints, which is the generally accepted joint thickness for structural clay tile construction.

CONSTRUCTION METHODS

In general, mortar for structural clay tile units is the same as that used for concrete block or brick. This means that a lime-portland cement type is used. For exterior walls, a type *M* or a type *S* mortar is used. Type *M* requires 1 part portland cement, a quarter part of hydrated lime and 3 parts of sand. Type *S*, which provides a bit more lateral strength, requires 1 part portland cement, a half part of hydrated lime and 4½ parts of sand.

For interior uses where a load bearing wall is not expected to receive a compressive stress greater than 100 pounds per square inch, type *O* mortar can be used. Type *O* mortar consists of 1 part portland cement, 2 parts of hydrated lime and 9 parts of sand. In most residential construction, this will provide sufficient strength. The load limit for structural clay tile walls is about 300 pounds per square inch, no matter the type of mortar used.

As with all materials used to make mortar, it will pay you to check for quality and cleanliness. Also, the amount of water used should be just enough to provide the needed plasticity. The less water used, the greater the final strength of the mortar. Mortar should not slide off the trowel, nor should it slip off the end of the tile when the tile is buttered before being set in place.

TOOLS

The tools used for clay tile masonry are essentially the same as for brick masonry. Included are a mason's brick chisel (Fig. 14-4), a plumb bob (Fig. 14-5) and a brick mason's hammer (Fig. 14-6). In addition, you'll need three trowels and a jointing tool. The trowels each do a specific job, one is used primarily for laying the bed joints, the second, with a rounder end or tip, is used to butter the ends of

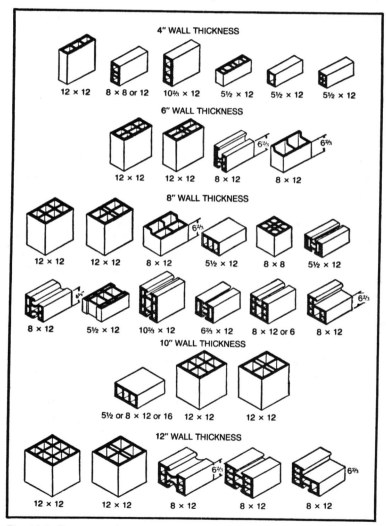

Fig. 14-3. Common configurations of structural clay wall tile.

Table 14-2. Hollow Clay Tile Wall Side Construction Material.

Wall thickness—tile size—wall area, sq ft.	4 inches—4 × 5 × 12		8 inches—8 × 5 × 12	
	Number of tile	Cu ft. mortar	Number of tile	Cu ft. mortar
1	2.1	.045	2.1	.09
10	21	.45	21	.9
100	210	4.5	210	9.0
200	420	9.0	420	18
300	630	13.5	630	27
400	840	18.0	840	36
500	1,050	22.5	1,050	45
600	1,260	27.0	1,260	54
700	1,470	31.5	1,470	63
800	1,680	36.0	1,680	72
900	1,890	40.5	1,890	81
1,000	2,100	45.0	2,100	90

*Quantities are based on ½-inch thick mortar joint.

the tile or brick and the smallest is used for pointing. The jointer is used to finish the joints and it is available in a variety of shapes in addition to that shown in Fig. 14-7.

You will also need a good level that is at least 2 feet long, but preferably 3 or 4 feet in length, and some mason's cord. A steel framing square will also prove handy for keeping corners square.

Just as with brick masonry work, a mortar board to hold mixed mortar and a mortar box for hand mixing mortar will be needed (Figs. 14-8 and 14-9).

Bonds

In most cases, structural clay tile will be laid with a running bond, though stacked bonds and other styles can be used if you prefer. Be sure to carefully align the cells of any wall that has plumbing or wiring. This can be done as easily using a running bond as with a stacked bond and the resulting wall is stronger. It is probably best to have most of the pipe or wiring already in place, but unconnected on its upper end. This makes it possible to slip the tiles right over the wire or pipe for nearly automatic alignment.

Joints

There is not a lot of difference in the mortar joints used for structural clay tile and those used for concrete block. The bed joint for end construction tile units is produced by spreading a 1-inch thick layer of mortar on the shell of the bed tile—not the webs.

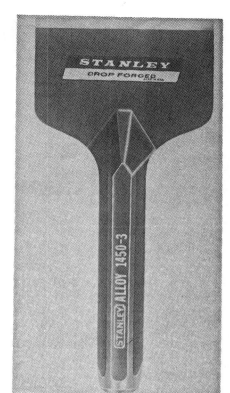

Fig. 14-4. Mason's chisel
(courtesy Stanley Tools).

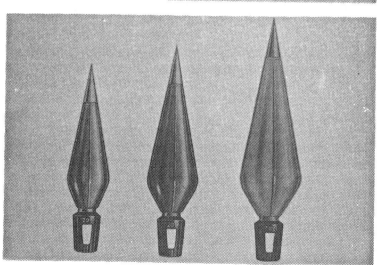

Fig. 14-5. Three sizes of plumb bobs (courtesy Stanley Tools).

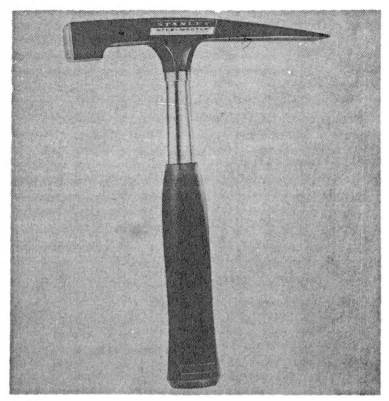

Fig. 14-6. A steel handled brick hammer (courtesy Stanley Tools)

Mortar is spread for a distance of 3 to 4 feet ahead of your tile laying. You might have to modify this somewhat, since the mortar could set up too quickly if you lay the tile slowly. Head joints are best staggered in a running bond style. No head joint should, match the one directly below it. Because the webs do not make contact, mortar spread on them is a waste of time and materials.

As Fig. 14-10 shows, end construction tile is well buttered with mortar before it is layed. Spread the mortar along both sides of the mortar bed. If enough mortar is used to make a good joint, some will squeeze out. The excess mortar is cut off with your trowel.

With structural clay tile units, head joints need not be (and sometimes cannot be) solid as with brick masonry. If the joint will be exposed to weather, it should be as well filled as possible.

Because of the size of most structural clay tile units, you'll probably need both hands to position them. The cells make them

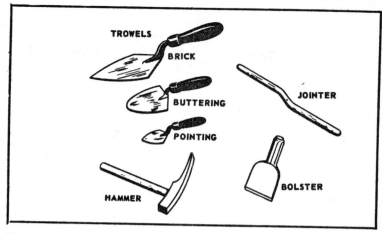

Fig. 14-7. Mason's tools.

lighter than a solid unit would be, but they are almost always much larger in size weight than single bricks. Mortar joints will be about one-half of an inch thick.

When laying the mortar for a bed joint for side construction, the mortar is spread to a length of 3 feet or so and about 1 inch deep. No furrow is needed, as is necessary with brick.

For side construction, head joints can be made one of two ways. With the first method (Fig. 14-11), as much mortar as will stick is placed on each edge of the tile and the tile is pushed into place in the mortar bed already run. Excess mortar is cut off with

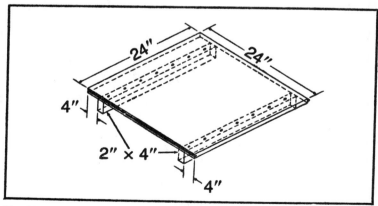

Fig. 14-8. An easily made and conveniently sized mortar board. Use exterior grade plywood at least ⅝ of an inch thick.

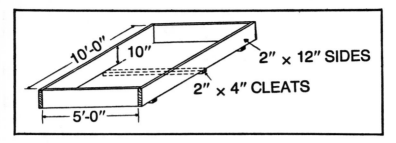

Fig. 14-9. You might prefer to down size this mortar box design since it is large enough to provide mortar for a professional. The mortar might set up before it is used, if the entire box is used. Cutting the length to 5 feet will help reduce waste.

the trowel. If you choose to use the second method, place as much mortar as will adhere to the edge of the already positioned tile, on the side on which you are standing. Then place as much mortar as will stick on the outside edge of the tile to be set in place. This tile is then pushed into its proper position and the excess mortar is cut off. (Fig. 14-12). Joints should be about one-half inch thick.

TILE BACKED WALL

To construct an 8-inch wall, using 4-inch wide structural clay tile as a backer for brick, use six stretcher courses between header

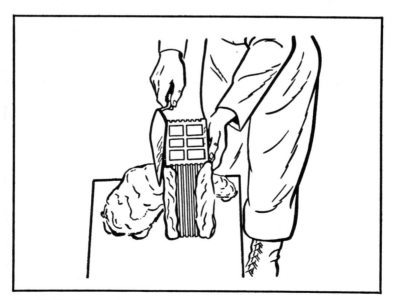

Fig. 14-10. Butter end construction tile before laying it.

Fig. 14-11. One method for making a head joint when using side construction hollow clay tile.

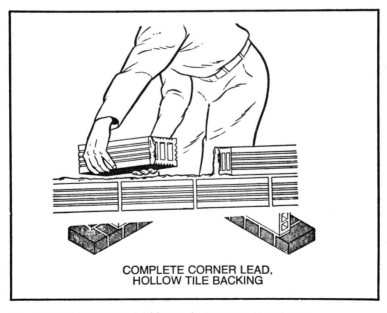

COMPLETE CORNER LEAD,
HOLLOW TILE BACKING

Fig. 14-12. A second method for producing a good head joint.

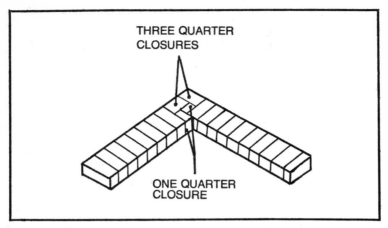

Fig. 14-13. The first course of the corner lead on a brick wall to be backed with hollow clay tile.

courses, and side construction tiles and methods. Generally, such a wall is built using 4 × 5 × 12 inch tiles, since the 5-inch height equals the height of two brick courses plus a one-half inch mortar joint.

Lay out the first course of the wall without mortar so that you can figure the number of bricks as well as tile that is needed for a single course. Each course of tile must be laid in a bed of mortar just thick enough to bring its top level with every second course of brick.

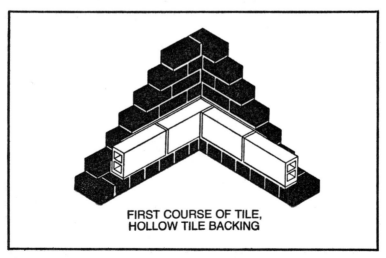

Fig. 14-14. All the brick in the corner lead is laid before the first course of hollow tile backing is placed.

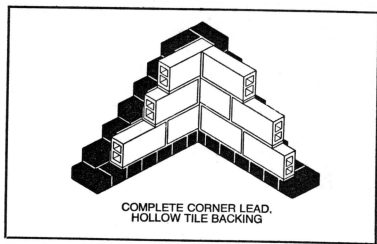

COMPLETE CORNER LEAD,
HOLLOW TILE BACKING

Fig. 14-15. A completed corner lead.

Therefore, you might have to adjust joint thickness from the usual one-half inch.

Note that the first course of the wall is entirely brick (a header course) and that three-quarter closures are used. Lay one more brick than you would lay starting a solid brick wall so that the tiles will fit on the inside of this course without overhanging the brick. Figures 14-13, 14-14 and 14-15 show the correct layout for the

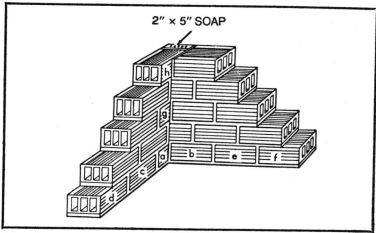

2" × 5" SOAP

Fig. 14-16. Using hollow clay tile as the only wall component requires this type of construction. It is a good idea to lay out a course without mortar first. This way joint size and needed cuts can be determined.

283

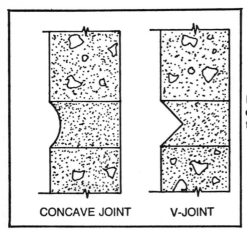

CONCAVE JOINT   V-JOINT

Fig. 14-17. Joints tooled in either of these manners are the most watertight.

corner leads. All bricks for the corner leads are laid before any tile is placed.

Joints are finished just as they would be with brick, using concave or V-shaped joints for exterior surfaces to prevent water penetration. Any style, from struck to raked to V or concave, can be used on interior walls.

CLAY TILE WALL

To construct an 8-inch wall entirely of structural clay tile, you start using 8 × 5 × 12-inch tiles, with 2 × 5 × 8-inch soap tiles used at the corners (Fig. 14-16). As the illustration shows, it is best to use a half lap bond. The first bed joint is laid on the footing or foundation wall, going just beyond the 3-foot mark to allow for the first course having at least three tiles laid in each corner lead. If you check Fig. 14-16, you will note that tile A is laid first, and B butted against it, with tile A projecting just six inches along the mortar bed to provide for the half lap joint. Corner tiles (B, G, H) must be of end construction style to keep from exposing open cells. A soap tile is a special, thin, end construction tile.

Once the corner leads are laid and the levels checked, run a mason's line for each course and lay the rest of the wall, continuing the half lap bond.

Joint finish follows as previously mentioned, with V or concave joints used for exterior surfaces (Fig. 14-17). Any joints which have not been properly filled during the construction of the wall should be pointed. Extra mortar is added and jointed as soon as possible in order to make sure the wall is as weathertight as possible.

15

Stone Masonry

If the cost of concrete block, brick or structural clay tile tends to make your wallet scream, there is another course to follow. Stone masonry is especially handy if you are building retaining walls in a yard or simply working for a decorative look in a garden or other outdoor area. Stone materials can also be used to build quite attractive barbecues, planters and other outdoor projects such as fireplaces and chimneys. In the case of barbecues and fireplaces, it is essential that the fire box be lined with firebrick. The reason is simple. Stones are generally larger than bricks and stones often have a great deal of moisture trapped inside them. Much depends on the type of stone, but most all will have an appreciable amount of moisture. This moisture will expand and could explode the stone under extreme heat.

CHOOSING MATERIALS

Suitable stone for most types of stone masonry can be found almost everywhere in the country. Types used include granite, limestone, sandstone and slate. Obviously, of these the most durable under extreme weather conditions is granite. But is also tends to be harder to locate than either sandstone or limestone and might (if quarried) cost as much or more than brick clay tile.

Tools

Most stone masonry requires the same tools as does brick masonry. In other words, trowels, a jointing tool, a bricklayer's

Table 15-1. Stone Masonry Mortar Mixes. Proportions by Volume.

Type of service	Cement	Hydrated lime	Mortar sand, in damp, loose condition
For ordinary service.	1—masonry cement*	----------	2¼ to 3.
	or 1—portland cement.	½ to 1¼	4½ to 6.
Subject to extremely heavy loads, violent winds, earthquakes, or severe frost action. Isolated piers.	1—masonry cement* plus 1—portland cement or 1—portland cement.	0 to ¼--	4½ to 6. 2¼ to 3.

hammer, a mason's chisel, mason's cord, a spade or shovel and a hoe. Also, a good level is useful, although stone masonry is not as neat as either concrete block or brick. The level can be a short one and a straightedge some 6- to 8-feet long can be made so that the level works over a fairly long surface, as there will be a fair number of irregularities in most stone walls. In addition, a plumb bob to keep the wall as close to plumb as possible and a carpenter's square for any 90 degree corners are needed.

In some cases, it is easier to use a power saw with a cut off wheel for masonry than it is to use a chisel or brick hammer. I've often taken the edge off a brick chisel while trying to cut granite and then finally switched to cut-off wheels whenever possible. If you use such a wheel, there are two precautions. Don't buy one with a plastic arbor insert, as the insert wears out within minutes. Always wear goggles (a good idea even with the chisel and a face mask. Cut-off wheels toss up a tremendous amount of very, very fine dust which cannot help but be bad for your lungs.

Mortar

Two types of mortar are considered equal to the job in stone masonry. One, for general service, requires 1 part of portland cement to a half part of hydrated lime to 4½ parts of sand. For walls or other structures which might need to withstand severe loads (very strong winds, or very sever frost action), use 1 part of portland cement, to a quarter part of hydrated lime, with only 3 parts of sand (Table 15-1).

Stone masonry can be either coursed or random. You can lay the stone in courses, similar to brick, or you can place the stones pretty much any way they'll fit into the wall. Really crude stone masonry is called random rubble, and the stones are of varying sizes and laid every which way. The only requirement is that each layer of stones must contain bonding stones of great enough size to extend the depth of the wall (Fig. 15-1).

Rubble masonry (with odd sized stones) need not be done in a totally random manner (Fig. 15-2). With careful stone selection, it can be coursed as shown in Fig. 15-3. Pick the stones that are closest to square or rectangular. Make sure you have a good variety of sizes. With all stone masonry, you will want a good selection of all stone sizes. If you don't have plenty of small stones to fill in large gaps the amount of mortar you will use will be excessive.

LAYING STONE

Stone masonry footings are usually made up of stone and the gaps are filled with mortar. This is not essential since a regular footing can be laid and the first course of the masonry started on that. However, it is generally cheaper to use large stones, placed below the frost line, and to fill in with mortar. Stones used for

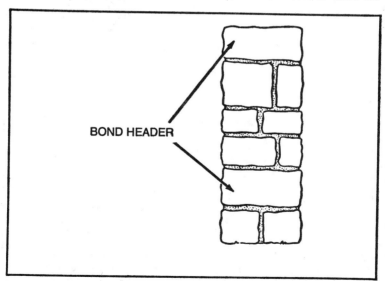

BOND HEADER

Fig. 15-1. Bond stones placed so as to strengthen the wall.

footings should be at least as long as the footing is to be wide. In general, stones should be laid on their broadest face to aid wall strength.

Most people feel that a rubble masonry wall looks best if the largest stones are used at the base and the size of the rock used gradually shrinks as the top of the wall is approached. In some cases, you might want to use stones arranged in a decorative pattern. This might preclude laying all the large stones down low, but in most cases walls will appear better balanced if you do use the largest stones at the base.

Like brick, some stone is absorbent and must be dampened before being placed in the wall. If this isn't done, the stone will draw off so much water that mortar hydration will not be complete and the wall will be weakened. To check to see if your particular type of stone is absorbent or non-absorbent, use the sprinkle test. Shake a few ounces of water on the stone. If it runs off, the stone is non-absorbent. If it soaks in, you'll need to wet the stone before laying it. Sprinkle with a garden hose until water runs from the pile. Allow to stand for a time before laying the first stones.

Select the stone so that spaces between each stone in the wall are as small as possible. Have enough small stones handy so that any unavoidable large gaps can be filled, in part, with small stones. Embed these well into mortar, of course.

Mortar bed joints for stone masonry don't follow hard and fast rules for thickness. One stone might need only a quarter of an inch of mortar to be firmly placed, while the stone next to it might require a joint varying from a quarter inch to a full inch. Just make sure there

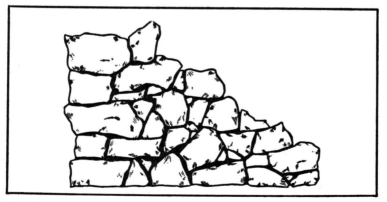

Fig. 15-2. An example of random rubble masonry, showing how small stones are used to reduce the waste of mortar.

288

Fig. 15-3. Coursed rubble masonry requires careful selection of stones for shape and size.

is enough mortar in each bed joint to completely fill all the space there is.

Head joints don't go quite as they do with brick. It is usually impossible to butter stone—especially if it has some odd contours. Head joints in masonry are generally "slushed" full of mortar. Make certain that the joint is well filled.

Cut stone masonry is similar to brick masonry in the joints used and the manner of making them. The method is often called *ashlar stone masonry*. A joint thickness of no more than half an inch is used and head joints go every which way.

To point stone masonry, you rake out the mortar in the joints to depths up to three-quarters of an inch. Many people, myself included, prefer a flush joint, or as near flush as is possible on such a rough surface, and one that has been pressed tight for water resistance. While rubble stone masonry is relatively inexpensive, there is no point in putting a lot of work into a job only to have water get into the joints, freeze, expand and crack the wall the first or second winter after it is erected. The best tool I've found for pressing stone masonry joints tight is a tongue depresser. Get quite a few. since the mortar wears them down rather quickly.

Easy Outdoor Projects

The projects in this chapter are attractive, uncomplicated, practical ideas for do-it-yourselfers. Before you actually begin construction, it is often a good idea to discuss the location of the project with your family. For example, you might want to build a sandbox that is in the line-of-sight of a specific window. Except for the very simplest projects, always check local building codes. Don't forget that these requirements are based on specific reasons such as the frost line for your area. You're going to be more than just disappointed if you build a concrete base for a wall and realize too late that it should have been deeper.

ESTIMATING MATERIALS

There are over 10,000 combinations of brick colors, textures and sizes. For outdoor projects, *SW* grade (Severe Weather) brick is recommended in most areas of the country. Table 16-1 and Table 16-2 can be used to estimate the amount of material you will need to build walls or pavement. You should always buy enough extra material to allow for some waste. This is especially true if a large number of bricks must be cut. Estimates for the projects in this chapter do not include waste unless specifically noted.

For most of these projects, using premixed mortar is a convenient and practical choice. It is slightly more expensive, but premixed mortar does save time and trouble. This is especially true for the smaller projects.

Table 16-1. Estimating Brick and Mortar.

Walls				
Brick Size	**Brick Per 100 sq. ft.**		**Cubic Feet or Mortar Per 1000 Brick**	
	⅜″ joint	½″ joint	⅜″ joint	½″ joint
3⅝ × 2¼ × 7⅝ 3¾ × 2¼ × 8	675 655	616	8.1 8.8	11.7
Pavements				
Brick Size	**Brick Per 100 sq. ft.**		**Cubic Feet of Mortar Per 100 Brick (includes ½″ bed)**	
			⅜″ joint	½″ joint
2¼ × 3¾ × 8 2¼ × 3⅝ × 7⅝ 1⅝ × 4 × 8	400 450 450		1.49 No Mortar Required	1.84

Tools

For most projects, a bricklayer's trowel, mason's string, a mason's hand level, a 2-foot hand level, a 4-foot hand level, a rubber mallet, a framing square and a broad-bladed brick chisel will be sufficient. Other tools can be improvised. A short length of three-quarter inch steel pipe can be substituted for a mason's pointing tool or a screed board can be made from an 8-foot long 2 × 4. The screed board is used to make a level surface across the pavement or other surface and to control the depth of a sand cushion. Make a 2- or 3-inch notch in the end of the board so that it will ride on guide boards.

A wooden tamper can also be made. Obtain a 2 × 6 or 2 × 8 section of 2-inch lumber. Cut it to 1 foot in length and nail a 2 × 2 on as a handle. Be sure to sand the edges to guard against splinter.

Table 16-2. Estimating Sand and Gravel.

Square feet per ton			
	1″ thick	2″ thick	4″ thick
sand gravel	276 229	138 116	69 58

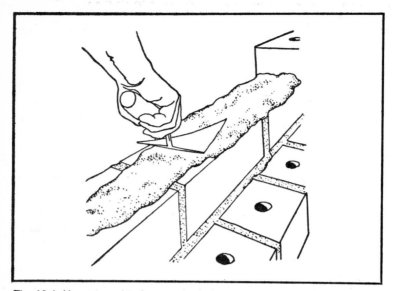
Fig. 16-1. Use a trowel to furrow a bed of mortar.

Site Preparation

These projects generally work best over a level concrete base. Excavate to the proper depth and prepare the concrete. Concrete is most easily handled from a wheelbarrow with a deep carrying base and nearly vertical sides. Use a pointed shovel and a garden hoe for manipulating the concrete. A rake does not work as well as a hoe when you are moving the concrete in the excavation. If you are working alone, remember that the concrete should season for at least one week. Slabs should be allowed to season two weeks or more. Spray thoroughly with water and cover the concrete with burlap or a cloth sheet. Dampen the cover every day and especially during dry, hot weather.

Workmanship

Before you start a project, it's often a good idea to lay the first few courses of brick in a practice run. This gives you a chance to recognize potential problems before they become major difficulties.

Some types of brick should be laid completely dry and others should be dampened first. There is a simple test you can do to find out which type you have. Draw a circle about the size of a quarter on one of the bricks. Using a medicine dropper, drop 20 drops of water inside the circle. Wait 90 seconds. If the water is visible, the bricks should be laid dry. If the water is not visible, the bricks should be laid damp—but not dripping wet. About 15 minutes before you begin construction, hose the pile of brick.

Unless otherwise noted, allow three-eighths of an inch between bricks for mortar joints for projects described in this chapter. Mix only a small amount of mortar to a consistency of soft mud. If it begins to stiffen, temper it by mixing in a small quantity of water.

Bricklayers refer to a *shoved joint* when describing the proper way to position a brick. Spread a bed of mortar and roughen the mortar surface by making a furrow with the point of the trowel (Fig. 16-1). Do not attempt to do this for more than three bricks at one time.

Next, "butter" one end of a brick with mortar and shove it into the mortar bed with a downward movement (Fig. 16-2). The top level should be level with the string line. When you are skilled enough to be able to position bricks without moving them once they are in place, you are making professional shoved joints.

Moisture is one of the biggest problems faced by masons. Unless the project is watertight—and only shoved joints make this possible—water that is trapped between the bricks will repeatedly freeze and thaw and eventually this will destroy the project.

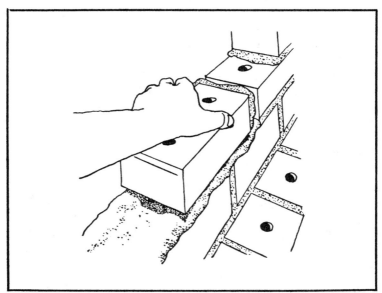

Fig. 16-2. Butter the brick and shove it into place.

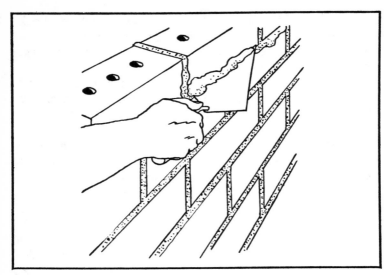

Fig. 16-3. Clip off and reuse excess mortar.

As you lay the bricks, use the edge of the trowel to cut off excess mortar (Fig. 16-3). The excess mortar can be returned to the mortarboard.

After the bricks have been positioned and the mortar is thumbprint hard, the joint must be tooled. Use the mason's pointing tool or a short length of three-quarter inch pipe to press each joint (Fig. 16-4). Work on the horizontal joints first and then the vertical ones to form hard, dense concave joints.

Bricks can be cut with a broad-bladed chisel. Tap the chisel with a hammer to score the brick along the line of the cut. Score on two surfaces of the brick. Next, point the chisel inward and strike a sharp blow with the hammer. A clean break should result.

BRICK EDGING

One of the easiest outdoor brick projects to make is brick edging placed around lawn and garden areas. Use a flat-bladed spade to dig a continuous cut about 4 inches deep and about 10 inches wide. Spread a layer of sand and place the bricks flat (Fig. 16-5). Slight curves can be made by slightly fanning the individual brick. For sharper curves (Fig. 16-6), cut a wedge-shaped brick. Weather changes can cause the brick edging to rise or sink. This can be countered by adjusting the level of gravel underneath the displaced bricks.

Tools

Flat-bladed Spade
Broom
Shovel
Rubber mallet

Material

32 solid brick per 10 feet of edging
1 ton of sand per 110 feet of edging (allows for waste)

STEPPING STONES

For best results, stepping stones should be placed about 4 inches apart. Before you begin to dig, determine how many steps you want and the location of each one. Then for each step, excavate a hole 4 inches deep and large enough to receive a 17½-inch redwood frame (outside dimensions). Position the frame so that it is level with the grass line (Figs. 16-7 and 16-8). Mix 1 part cement to 3 parts sand and spread 1 inch to 1½ inches of this mixture (dry) into the hole and tamp. Position several of the bricks to determine if they will be flush with the top of the frame. Adjust the sand cushion as required and lay the bricks in the pattern you have selected.

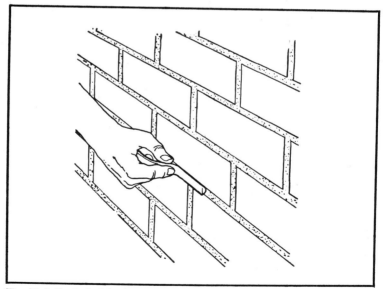

Fig. 16-4. Tool all of the joints to obtain best weather resistance.

Fig. 16-5. Place the bricks flat and tight against each other.

Tools

Shovel
Garden hoe
2-foot hand level
Wooden tamper
Framing square

Material

For 10 square that are 16″ × 16″:
80 solid brick units 3 ¾″ × 2¼″ × 8″
10 lengths of redwood plank 1″ × 4″ × 6′

Fig. 16-6. Brick edging can be curved.

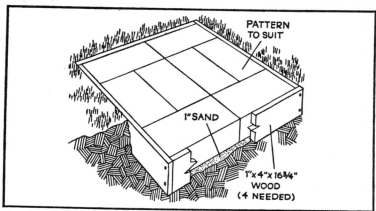

Fig. 16-7. The frame for a stepping stone should be level with the grass line.

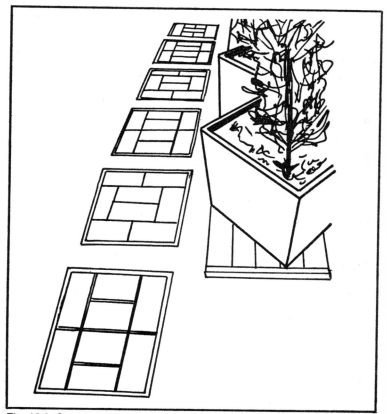

Fig. 16-8. Stepping stones make an attractive addition to a lawn.

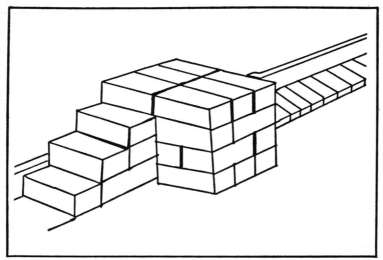

Fig. 16-9. A pedestal is an easy project to build.

Galvanized box nails (five-penny)
3 cubic feet of sand (about 260 pounds)
1 bag (94 pounds) of portland cement

PEDESTAL

Brick pedestals can be nice settings for flower arrangements. Because no mortar is used for this design, it is essential that a level, firm base be available. A concrete slab, 16 × 12 inches square and about 4 inches thick, is best but a gravel base can be used. Be sure to excavate below the frost line. Position the bricks as shown in Fig. 16-9. Lay as many of the courses as possible so that cracks between units are not lined up.

Tools

Shovel
Wheelbarrow
Garden hoe
Wooden flat
Trowel

Material

30 solid bricks
One-half cubic foot of concrete (1 bag premixed concrete)

PLANTER

The planter shown in Fig. 16-10 is not difficult to construct. If you are more ambitious, you might want to change the size and shape. This can easily be done, but be sure to adjust the amount of material to match your plans. This type of planter is best built on a concrete base. Lay the bricks in place with mortar as shown in Fig. 16-11. Use nylon rope or plastic tubes to add "weep holes" in the first course of brick Let the planter season for about one week. Be sure to spread a layer of gravel in the bottom of the planter to insure proper drainage.

Tools

Shovel
Wheelbarrow

Fig. 16-10. This type of planter should be built on a concrete base.

Fig. 16-11. The layout of a small planter.

Garden hose
Garden hoe
Hand brush
Bricklayer's trowel
4-foot hand level
Bottomless wooden box
Framing square
Hammer
Pointing tool
Broad-bladed brick chisel

Material

about 15 solid bricks for each planter

premixed mortar, one 60-pound bag will be enough for 3 planter (including waste)

nylon rope or plastic tubing

SANDBOX

A sandbox is not particularly difficult to construct and it will give your children a fun place to play (Fig. 16-12). Excavate for the project as indicated in Fig. 16-13. Install the wooden frame and the 1 × 4 edging. Mix 3 parts sand to 1 part cement and spread over the ground around the sandbox. Tamp and grade as necessary to obtain a smooth and level surface. Position the bricks and adjust as necessary to obtain a level, tightly packed surface. For an alternate design, see Fig. 16-14.

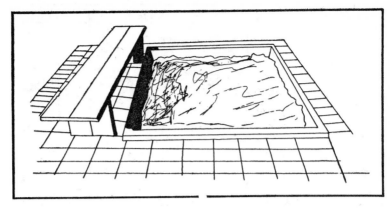

Fig. 16-12. A sandbox with a wooden frame.

Tools

Shovel
Wheelbarrow
Garden hose
Garden rake
Broom
Mason's string
Hand level
2-foot or 3-foot hand level
Brick chisel

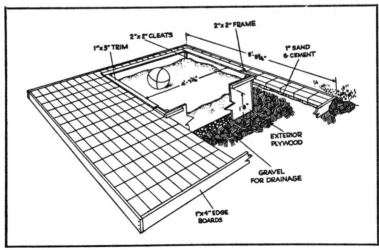

Fig. 16-13. Sandbox dimensions.

Fig. 16-14. A sandbox is a good way for a novice bricklayer to gain experience.

Wooden tamper
Wooden screed board
240 solid bricks, 3 ¾″ × 2¼ × 8″ or 4″ × 8 × 1 ⅝″

About 5 cubic feet of damp, loose sand (¼ ton). This does not include sand for the sandbox.

1¼ cubic feet of cement of 1½ bags of portland cement
Galvanized nails

⅝″ exterior grade, rough sawn plywood siding (coated with preservative); cut as follows:

4 lengths of 20″ × 55 ⅜″
4 lengths of 2″ × 2″ × 7′
4 lengths of 1″ × 4″ × 9′
2 lengths of 1″ × 3″ × 10′

STEPS

Concrete steps can be covered with brick (Fig. 16-15) to improve the appearance and durability of entryways. This project takes careful planning and workmanship. Before you mix the mor-

Fig. 16-15. Brick covered concrete steps.

tar, lay out the position of the bricks in a "dry run." Remember to allow for three-eighths of an inch between bricks for the mortar. If you will be cutting a great many bricks, it might be convenient to rent a brick cutting machine. Be sure that the concrete steps have an adequate forward slope. To insure proper drainage, lay the bricks so that they slope forward slightly (Fig. 16-16). Be particularily careful to tool all mortar joints inward. This will protect the steps from water penetration.

Fig. 16-16. Slope the bricks to obtain proper drainage.

Tools

Shovel
Wheelbarrow
Garden hose
Garden hoe
Hand brush
Brick cutter (optional)
Trowel
2-foot hand level
Bottomless wooden box
Pointing tool
Broad-bladed brick chisel
Hammer

Material

Allow 4.1 units per square foot of tread area and 2.75 units per foot of length of the risers for brick that is 4″ × 8″ × 1 ⅝″
Mortar (consult estimating table)

SCREEN

A brick screen can be attractive as well as functional. It can be used to keep trash cans (Fig. 16-17), utility installations and carports out of sight (Fig. 16-18). Some building regulations do not permit brick screens. Check the local codes to determine the allowable height for a screen and the proper depth of the concrete footing. Before you begin construction, study Figs. 16-19 and 16-20 very carefully. It is a good idea to have some experience with a less complicated project before you attempt this one.

Excavate the footing to below the frost line. Use a 2-foot or 4-foot hand level and an 8-foot long 2 × 4 to make sure the bottom of the trench is as level as possible. Dig the footing no wider than necessary. Using the sides of the trench for forming will enable you to by-pass the cost and labor of extra material.

The corners of the screen must be constructed as plumb as possible. Laying out the first few courses in a "dry run" will give you a chance to spot potential problems.

After you have constructed the solid portion of the wall, begin the pierced-pattern section of the screen. Cut a brick to 6 inches in length and lay it in place (see A of Fig. 16-20). Always position the cut bricks so that the cut portion is visible only from the outside of the screen. Next, cut a brick in half and position it so that it is centered on the joint where two bricks come together (see B of Fig.

16-20). Start the next course with a half brick (see C of Fig. 16-20). Follow with a whole brick that will span to the middle of the cut brick positioned one course below (see D of Fig. 16-20). Continue the remainder of the course and repeat the pattern.

Avoid putting pressure on the screen as you build it. This type of wall is not as sturdy as a solid wall and, of course, the joints will not be completely hardened immediately.

Tools

Shovel
Wheelbarrow
Garden hose

Fig. 16-17. The screen design can be used in building a storage bin.

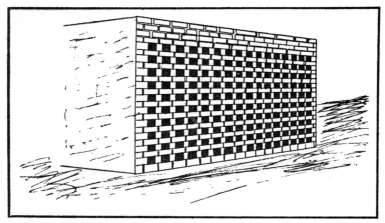

Fig. 16-18. Solid and cut bricks form a screen pattern.

Garden rake
Hammer
Broom (short, stiff bristles)
Garden hoe
Brick cutting equipment (optional)
Bricklayer's trowel
Mason's string
Mason's had level
2-foot and 4-foot hand levels
Framing square
Mason's pointing tool
Broad-bladed brick chisel
Wooden float

Material

Allow 4.4 units per square foot of screen; 6.55 units per square foot of solid wall for brick that is 3 ¾″ × 2¼ × 8″ (allow 5 percent to 25 percent for waste)

Mortar (consult estimating table)

.90 cubic feet of concrete per lineal foot for 8′ × 16″ footing
.66 cubic feet of concrete per lineal foot for 8″ × 12″ footing

RETAINING WALL

A brick retaining wall can protect you property from erosion and enhance the appearance of your property. Although this wall is designed to be built no more than 3 feet high, it is a project that requires excellent workmanship and a good deal of ambition. Plan ahead, study Figs. 16-21, 16-22 and 16-23 very carefully and remember to comply with local building codes.

Excavate for the footing. Lay the bottom reinforcing bars on some loose bricks. Wire the vertical bar to the bottom bar and prop in place. Insert the remaining bars into the top of the footing as the concrete is poured.

Allow the concrete footing to season for at least one week. Then lay the brick using patience and your best shoved joint technique. Be sure to insert the prefabricated steel joint reinforcement as indicated in Fig. 16-21. Position sections of 1 inch diameter plastic tubing for "weep holes" every 4 feet along the wall.

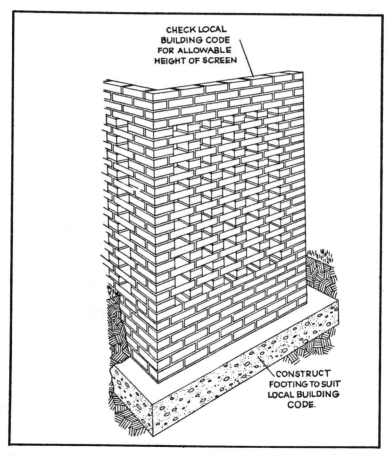

CHECK LOCAL
BUILDING CODE
FOR ALLOWABLE
HEIGHT OF SCREEN

CONSTRUCT
FOOTING TO SUIT
LOCAL BUILDING
CODE.

Fig. 16-19. Study this pattern closely before you begin construction of a screen.

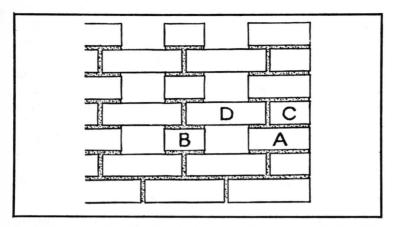

Fig. 16-20. The basic pattern for a brick screen. A brick is cut to 6 inches (A). Two bricks are joined (B). A new course starts with a half brick (C). A whole brick spans to the middle of the course below (D).

After the last horizontal course of bricks has been laid, pour grout (mortar which has been thinned with water) in the gap between the bricks. This will bond the reinforcing bars. Cap the wall with a solid row of bricks laid on edge.

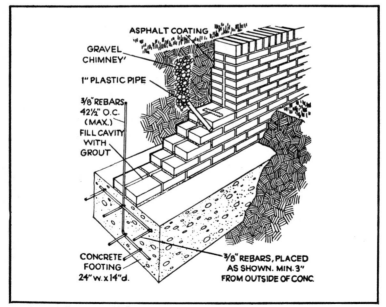

ASPHALT COATING
GRAVEL CHIMNEY'
1" PLASTIC PIPE
3/8" REBARS 42½" O.C. (MAX.)
FILL CAVITY WITH GROUT
CONCRETE FOOTING 24" w. x 14"d.
3/8" REBARS, PLACED AS SHOWN. MIN. 3" FROM OUTSIDE OF CONC.

Fig. 16-21. Dimensions of a retaining wall.

Fig. 16-22. Each course of the retaining wall must be plumb.

Coat the side of the wall that faces the earth with a brushed on asphalt coating to make the wall watertight. Place a french drain of gravel behind the wall and down to the level of the weep holes.

Tools

Shovel
Wheelbarrow
Garden hose
Garden rake
Hammer
Broom (short, stiff bristles)
Bricklayer's trowel
Mason's string
Mason's hand level
2-foot and 4-foot hand levels
Framing square
Mason's pointing tool
Wooden float

Fig. 16-23. Building a retaining wall requires excellent workmanship.

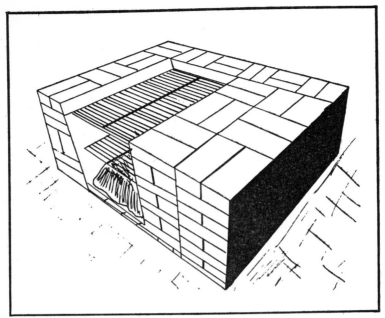

Fig. 16-24. A mortarless barbecue is easy to build.

Material

 1310 brick 3 ¾″ × 2¼″ × 8′ per 100 square feet of wall
 4.4 brick 3 ¾″ × 2¼ × 8″ per foot of wall length
 20 cubic feet of mortar per 100 square feet of wall
 2.33 cubic feet of concrete per foot of wall length
 Three-eighth inch steel reinforcing bars, 52″ long, bent 9″ from one end at a 90 degree angle—one for every 3½ feet of wall length
 Two 10-foot pieces of prefabricated joint reinforcement for 8″ wide wall for every 9½ feet of wall length
 Three-eighth inch steel reinforcing bars, 18″ long, one for every 3½ feet of wall length
 One-half inch steel bars for length of footing (allow for 10″ lapping splice)
 Plastic tubing
 Asphalt coating

MORTARLESS BARBECUE

 Because this barbecue is constructed without mortar, it is essential that the building site be absolutely level. A concrete slab

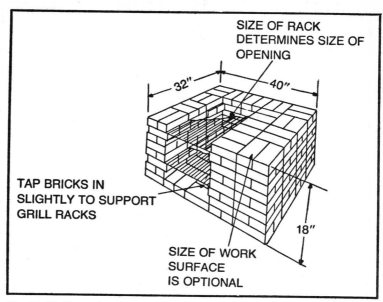

Fig. 16-25. Dimensions of a mortarless barbecue.

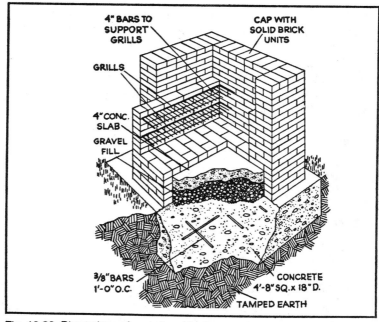

Fig. 16-26. Dimensions of a mortared barbecue.

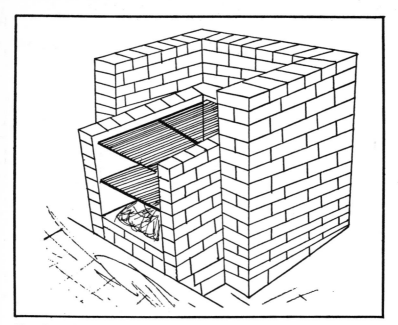

Fig. 16-27. A mortared barbecue.

provides the ideal surface. Be sure to obtain the grill racks before you actually begin construction. It is much easier to plan a barbecue around a grill than it is to buy a grill that will fit once the project is completed.

Study Figs. 16-24 and 16-25 before you begin. Using a hammer and a 2 × 4, carefully tap in the bricks that support the grill racks. Hold the 2 × 4 edgewise against the bricks as you tap them.

Tools

> 2-foot hand level
> Hammer
> 2 × 4

Material

> 236 solid bricks, 3 ¾″ × 2¼ × 8″
> Two grill racks

MORTARED BARBECUE

Choose the location for the barbecue with the prevailing winds in mind. Study Figs. 16-26 and 16-27 before you begin work.

Excavate the site and pour the concrete. Be sure to place the reinforcing bars as shown in Fig. 16-26. Position the bars in a grid pattern and support them with bricks so that the grid is approximately in the center of the concrete. The bars can be wired together to form a unit.

Outline the barbecue on the concrete slab. Lay a few courses of brick without mortar to determine if the pattern works. Remember to allow for one-half inch mortar joints.

Begin laying the corner first. The bottom bricks are bonded with mortar to the concrete slab. Build three or four courses at the corners and the fill in the walls. Use a hand level to frequently check the walls for plumb. As you build to the height of the grills, insert

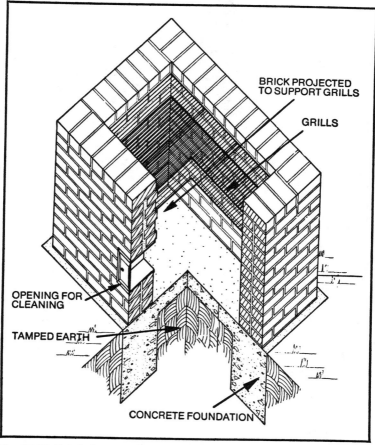

BRICK PROJECTED
TO SUPPORT GRILLS

GRILLS

OPENING FOR
CLEANING

TAMPED EARTH

CONCRETE FOUNDATION

Fig. 16-28. Details for a small cooking grill.

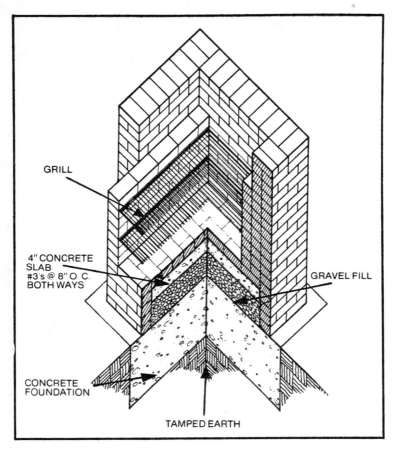

GRILL

4" CONCRETE
SLAB
#3's @ 8" O.C.
BOTH WAYS

GRAVEL FILL

CONCRETE
FOUNDATION

TAMPED EARTH

Fig. 16-29. Construction details for a mortared barbecue.

4-inch sections of reinforcing bars for grill supports (Figs. 16-26, 16-28 and 16-29).

Use a trowel to clip off excess mortar. When the mortar becomes thumbprint hard, use a pointing tool to work the joints.

Tools

Hammer
Mason's string
Trowel
Hand level
Brick chisel
Chalk

Wooden float
Hand brush

Material

450 cored bricks, 3 ¾" × 2¼" × 8"
75 solid bricks, 3 ¾" × 2¼" 8"
6 cubic feet of mortar
27 cubic feet of concrete for the foundation
1 2/3 cubic feet of concrete for the hearthslab
20 reinforcing bars, three-eights of an inch in diameter and of
the following lengths:

     5 bars 18" long
     3 bars 32" long
     12 bars 4" long

17

A Brick Patio and Planter

There are some general methods in mind before you begin to build the brick patio and planter described in this chapter. If you are using mortarless brick paving, remember that it has a tendency to spread at the edges if some sort of frame is not placed around the perimeter of the patio. This frame can be made of 2-inch wood (Fig. 17-1) or of a course of brick set in concrete or mortar (Fig. 17-2). If the edging is positioned prior to laying the brick surface, the edging can serve as a guide for drainage slopes.

Without proper drainage, brick paving can be subjected to excessive moisture that might result in the growth of fungi or molds and the disintegration caused by repeated freezing and thawing. To counter such potential problems, be sure to slightly slope the patio to one side or from the center at a rate of one-eighth to one-quarter of an inch per foot. Slope the paving away from buildings, walls or areas that will retain the flow of water.

Another potential problem can result from the upward capillary travel of moisture from soil that is heavily laden with soluble salts. Such salts can cause a powdery substance or stains to form on the brick paving. If the soil in your area contains soluble salts, install a "capillary break" of gravel (Fig. 17-3) to prevent the upward flow of moisture. This layer of gravel can be graded to obtain the proper slope and placement of the bricks. A layer of sand can be substituted for gravel if soluble salts are not present in the base soil. Do not place a layer of sand over a layer of gravel since the sand will

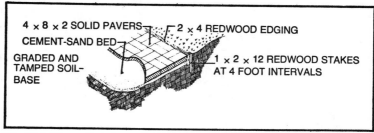

Fig. 17-1. The patio is framed with 2-inch redwood.

eventually sift into the gravel and cause uneven settlement of the brick paving.

A patio with a concrete slab base and mortared joint paving is the best way to counter the problem of ground swell caused by freeze/thaw cycles. However, if ground swell is not a problem in your area, a patio with mortared joints can easily be constructed without a concrete slab.

First, grade and tamp the subsoil to eliminate soft spots. Carefully screed a 1- to 2-inch cushion of sand over the subsoil. Lay the bricks in place with one-half of an inch of mortar between each unit (when using 3¾″ × 8″ × 2¼″ bricks). Be sure to sweep the paving free of dry mortar since portland cement will stain brick if it is allowed to remain on the surface. Keep the paving damp for two or three days.

To construct a patio without mortar joints, using 4″ × 8″ pavers, screed and tamp a 1. to 2-inch bed of sand over the subsoil. Next, roll out sections of 15-pound roofing felt and place these sections over the sand. Lay the bricks on the felt. The felt stabilizes the base and helps prevent the growth of weeds and grass between brick joints. After all the bricks have been layed, sweep dry sand into the joints.

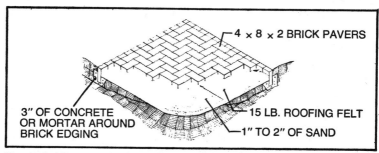

Fig. 17-2. Mortarless brick paving on a cushion of sand.

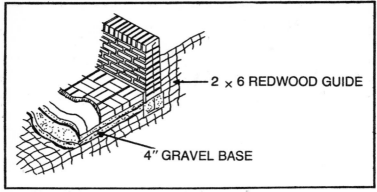

Fig. 17-3. A patio and planter with a cutaway showing a gravel base.

The patio layout shown in Fig. 17-4 is designed to be built in four sections to form a continuous surface around a large brick planter. Each section is approximately 6 × 14 feet and the completed patio is slightly over 20 square feet.

The project can be completed over a few weekends or during a short vacation if the bricks are laid on a cushion of sand and cement. In climates where freeze thaw cycles are common, the patio should be constructed on a concrete foundation.

LAYOUT AND EXCAVATION

Use stakes and string to outline the patio and planter. At the center of the line between stakes, suspend a line level. Remember to allow for a slope of one-eighth to one-quarter inch per foot. The stakes should be marked in the direction the surface will drain. these marks are also used as excavation guides and later they can be used as a guide for laying the bricks.

The outside dimensions for the patio are 6 feet 3½ inches × 14 feet 4 inches. Allow 1½ inches for redwood framing and an additional half inch to 1 inch for fit. The planter is 8 feet 1¼ inches square. See Figure 17-4.

Excavate 3½ inches deep for the patio section. For the planter, excavate a trench 3½ inches deep and 11 inches wide (form allowance) around the perimeter. Do not excavate the area within the planter.

During excavation, take measurements and frequently check the marks on the stakes. Stretch a line across the excavation and measure the depth, but don't be overly concerned with minor bumps and depressions. If you find that you have excavated too much at some point, fill in that area and tamp firmly. Soft spots could cause the paving to settle unevenly.

Build forms for the planter foundation with 2 × 4s. Don't be tempted to use the redwood framing since the redwood would probably be stained by the concrete. At 4-foot intervals, stake the forms into place with 1 × 2s driven securely into the ground. To make removal of the forms easier after the foundation has set, coat the inside of the boards with crankcase oil. After the forms are in place, check each side and the corners to make sure they are level. Using a garden hose, thoroughly soak the excavation within the form. Wait until the surface water disappears before pouring the concrete. When concrete is poured on dry soil, moisture is drawn out too rapidly. This can cause an improper cure that could result in a weak foundation.

Place 1 part cement, 2½ parts sand and 3 parts gravel into a wheelbarrow and mix with a shovel. Blend the contents thoroughly. Form a hollow in the center of the mixture and add a small amount of water. Use a garden hoe or a shovel to thoroughly mix the contents. Repeat this procedure, using approximately 3 quarts of water for each shovelful of cement. Too much water will weaken the concrete and too little water will make portions of mix too dry to set up properly.

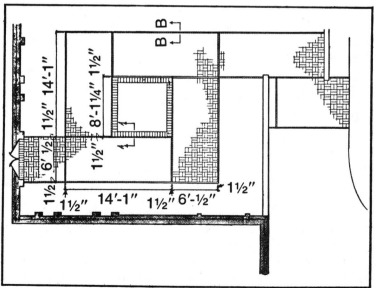

Fig. 17-4. A patio and planter layout.

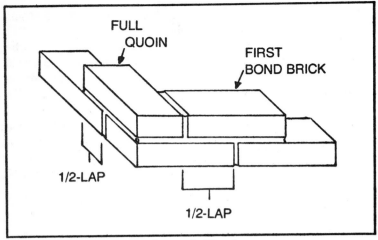

Fig. 17-5. Corner construction for one-half bond.

Shovel the concrete mix into the forms until the concrete is level with the top of the forms. Tamp the surface until level and fill in any low spots. The concrete should set for at least one hour. During this time, hose off the wheelbarrow and other tools. As the concrete hardens, use a stiff bristled broom to roughen the surface. This will create a bonding surface for the bricks.

To obtain proper curing, wet the concrete and cover it with burlap or straw in order to retard evaporation. In warm weather, keep the foundation damp for at least three days. In cooler weather, keep the foundation damp for as much as a week. After three or four days, you can begin laying brick. However, the forms should remain in place for at least one week.

THE PLANTER

The height of the planter is 13¾ inches and 5 courses of brick are used. Double rows of running bond are used for the first four courses. Each brick is offset half the length of the one next to it and the one below it. The top course is laid sideways across the course beneath it. This ties the rows together (Fig. 17-5).

Before you permanently set the bricks in place, position a few courses without mortar in order to check the accuracy of the measurements. Remember to allow for approximately one-half of an inch between each course for mortar joints. It should not be necessary to cut any bricks for the planter. If everything fits properly, spray the brick pile with water about one hour before you begin laying brick. The bricks should be damp but not dripping wet. Too much water will dilute the mortar and cause it to slip off the bricks when placed.

To form mortar, mix 1 part portland cement, one-half part hydrated lime and 4½ parts sand. Do not mix more than you can use in half an hour. Mortar should have the consistency of soft mud that slides easily from the shovel. If the mortar becomes too stiff, add a small quantity of water and mix thoroughly. Dampen the foundation lightly before you begin to lay the bricks.

The finished wall will have a thickness of 2 courses of brick—plus a one-half inch mortar joint. This is equal to the length of one brick.

Build the corners (Figs. 17-5 and 17-6) first by laying the two courses four rows high and dovetailed at right angles. Starting with the inside course, use a trowel to lay a half inch of mortar on the foundation, about the length of two bricks. As you gain experience you will be able to lay three or more bricks at one time. Furrow the mortar bed with the trowel point. Butter the end of a brick and set it in place. Clip off excess mortar and return it to the mortar board. Butter the end of another brick and position it against the first. Complete the inside course and then build the outside row, buttering the edges and sides to bond the two courses.

Use a straightedge and level to plumb the corners. After you finish all four corners, push nails into the soft mortar of the second course at each corner. Run a length of string between corners to

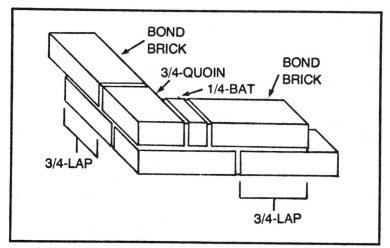

Fig. 17-6. Corner construction for three-quarter bond.

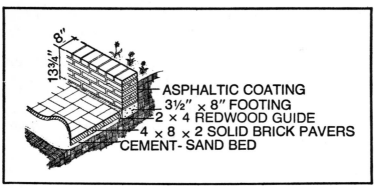

8"
13¾"

ASPHALTIC COATING
3½" × 8" FOOTING
2 × 4 REDWOOD GUIDE
4 × 8 × 2 SOLID BRICK PAVERS
CEMENT- SAND BED

Fig. 17-7. The top course of the planter is laid lengthwise.

serve as a guide while you lay the remainder of the bricks. Use a level and straightedge frequently to keep the walls plumb.

After the first four courses are completed, set the top course in place. Techniques for laying mortar and positioning bricks are the same except that the top course is laid lengthwise (Fig. 17-7).

Tool the joints with a pointing trowel or a section of three-quarter inch steel pipe to produce smooth concave surfaces. During hot weather, wet the brickwork occasionally the first few weeks after construction to allow the mortar to cure slowly. If mortar stains remain on the brick, muriatic (hydrochloric) acid diluted with 9 parts water can be used to remove the stains. For this job, wear rubber gloves and protection for your eyes. First, soak the brickwork with water and then scrub on the solution with a stiff bristled brush. Rinse the scrubbed area at once with water. After the mortar has cured, waterproof the inside of the planter with an asphalt coating.

THE PATIO FOUNDATION

The first step in building the patio foundation is to assemble the frame of redwood 2 × 4s. The inside dimensions should be 6' ½" × 14' 1". Check the measurements before you drive two 8d galvanized nails into each corner of the frame (Fig. 17-4).

Position the frame in the excavation. At approximately 4-foot intervals, drive 1 × 2 redwood stakes into the ground outside the frame. The corners of the frame should be square. Check the top edges of the frame against the marks on the grading stakes that were used to guide the excavation. The top of the frame and the top of the stakes should be even with the top of the finished paving. Nail the frame to the stakes (see Fig. 17-8).

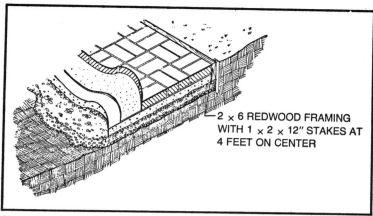

Fig. 17-8. Mortared paving and redwood edging.

Shovel sand into the center of the patio excavation. Leave a few shovelfuls of sand for later use in filling joints. Add one sack of cement at a time and thoroughly mix it with the sand. If the sand is at all damp, the mixture will begin to set. Therefore, you should spread the mixture with a rake as quickly as possible to a depth of about 1¾ inches.

Use an 8-foot long, 2 × 4 straightedge to screed the surface. The straightedge can be notched with the ends resting on the redwood frame (Fig. 17-9). Cut the notches one foot from the ends and 1⅞ inches deep. Draw the straightedge across the frame to level the patio bed. Fill in the low spots and repeat the screeding procedure until the bed is level.

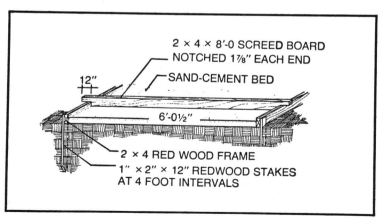

Fig. 17-9. Guide frame and screed board assembly.

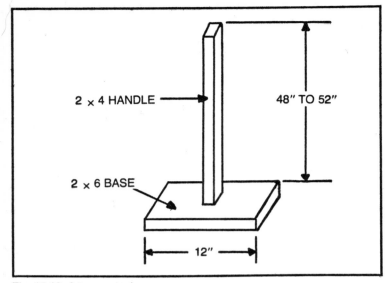

Fig. 17-10. A tamper tool.

Tamp the patio bed so that it is firm. A tamper can be built by nailing a short section of 2 × 6 to a 2 × 4 handle (Fig. 17-10). Fill in any low spots and repeat the screeding procedure.

LAYING THE MORTARLESS PAVING

Once the patio bed is prepared, begin laying bricks from the frame edge inward. Start at one corner and position the bricks tightly against the others in the pattern you prefer (Figs. 17-7 and 17-11). Stand on the laid bricks as you work rather than on the sand/cement surface. Use a mallet to firmly tap each brick into place. If the base has been properly prepared, the bricks should be level and each course should fit easily into the redwood frame.

After the last brick is placed, fill the joints with sand. Use a broom to sweep the surface clean. Wet the joints with a fine spray of water to compact the sand. Be very careful not to wash out the sand. All joints should be flush with the bricks. It might be necessary to refill some joints with sand and spray again. The patio is now ready for use.

BUILDING A PATIO ON CONCRETE

In a climate where winters are severe, the patio should be built on a concrete slab and the brick paving should be set in mortar.

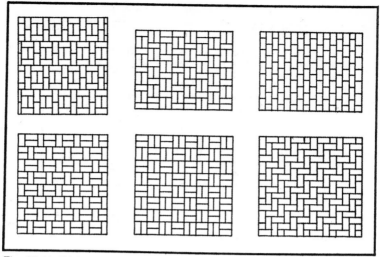

Fig. 17-11. Brick paving patterns.

Because mortar joints are used, building brick rather than paving brick is used for the patio.

The planter is constructed as previously detailed, except that the foundation is dug deeper. Check local building codes to determine the frost line your area. The overall dimensions of the planter are the same. However, the excavation of the patio increases to 6' 7" × 15' ½" to allow for half inch mortar joints and the guide frame. Lay out the section with stakes as previously described and excavate to a depth of about 9¾ inches.

The concrete slab is 6' 4½" × 14' 10½". Build forms from plywood or 2-inch lumber and position the forms in the excavation. The top of the poured concrete slab should be 2¾ inches below the top of the finished brick paving. Support the forms with stakes driven into the ground at 4-foot intervals. Refer to planter foundation details described earlier in this chapter.

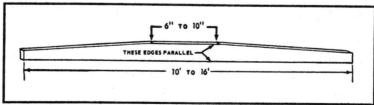

Fig. 17-12. A mason's straightedge.

Table 17-1. Concrete in Cubic Feet.

Col. 1 Footing Length (feet)	Col. 2 3½″ × 8″ Footing (cu. ft.)	Col. 3 8″ × 12″ Footing (cu. ft.)	Col. 4 Patio Area (sq. ft.)	Col. 5 3″ concrete slab (cu. ft.)	Col. 6 1″ Bed (cu. ft.)	Col. 7 2″ Bed (cu. ft.)
1	.19	.67	10	2.5	.83	1.66
2	.39	1.33	20	5.0	1.66	3.32
3	.58	2.00	30	7.5	2.49	4.98
4	.78	2.66	40	10.0	3.32	6.64
5	.97	3.33	50	12.5	4.15	8.30
6	1.16	4.00	60	15.0	4.98	9.96
7	1.36	4.66	70	17.5	5.81	11.62
8	1.55	5.33	80	20.0	6.64	13.28
9	1.75	6.00	90	22.5	7.47	14.94
10	1.94	6.66	100	25.0	8.30	16.60
40.5	7.86	27.00	108	27.0	8.96	17.93
139.2	27.00	92.69	150	37.5	12.45	24.90

Lay a 4-inch deep gravel base inside the foundation forms. Consult Table 17-1 to estimate the amount of concrete you will need. A 3-inch slab of concrete for each section requires approximately 1 cubic yard of concrete. Probably the most convenient method is to order ready-mixed concrete delivered by truck. Don't order more than 3 cubic yards at one time and be sure to have someone at the site to help you.

If you want to mix the concrete yourself, rent a mixing machine. Mixing by hand is very strenuous and time consuming labor. Refer to Tables 17-2 and 17-3 for the quantities of cement, sand, gravel, brick and mortar needed to construct the patio and planter.

The amount of cement you can mix at one time depends on the size of the machine you rent. Put 1 part cement and 2½ parts sand in the machine and mix thoroughly. Next, mix in 3 parts gravel. Add water and tumble for about 3 minutes. Pour the concrete into a wheelbarrow and then into the excavation.

Spread the concrete throughout the foundation and use a straightedge (Fig. 17-12) to obtain a smooth level surface that is even with the top of the forms. Use a wooden float (Figs. 17-13 and 17-14) to work out any irregular spots, but don't overwork the concrete. Overworking could result in a less durable foundation.

After the concrete sets for an hour, use a stiff bristled broom to roughen the surface. This will provide a bonding surface for the brick mortar. Water the concrete and cover it with burlap or straw

Fig. 17-13. A long handle wood float.

Table 17-2. Material List.

Number of 6′ × 14′ Patio Sections

Patio Material		1	2	3	4
Brick Pavers, 4″ × 8″ × 2″ (includes 5% waste)	=	400	800	1200	1600
Portland Cement (94 lb. sacks)	=	4	8	12	16
Sand, damp & loose (pounds)	=	1090	2180	3270	4360
Redwood framing, 2″ × 4″ × 8 ft.1	=	6	12	18	24
Redwood stakes, 1″ × 2″ × 12″	=	10	18	26	34
Nails, 2 pounds of 8d galvanized common					

Planter Wall Material

		Footings		Mortar		Total
Face Brick, 3¾″ × 2¼″ × 8″: 440 units (includes 5% waste)						
Portland Cement (94 lb. sacks)	=	1⅜	+	1¾	=	3⅛
Sand, damp & loose (pounds)	=	287	+	696	=	983
Hydrated Lime (50 lb. sacks)	=	—	+	¾	=	¾
Old lumber for temporary footing forms						

Number of 6′ × 14′ Patio Sections

Mortared Paving on Concrete Slab		1	2	3	4
Standard Brick, 3¾″ × 2¼″ × 8″ (includes 5% waste)	=	400	800	1200	1600
Portland Cement (94 lb. sacks)	=	3	6	9	12
Hydrated Lime (50 lb. sacks)	=	⅔	1⅓	2	2⅔
Sand, damp & loose (pounds)	=	772	1544	2316	3088
Redwood framming, 2″ × 6″ × 8 ft.1	=	6	12	18	24
Redwood stakes, 1″ × 2″ × 12″	=	10	20	30	40
3″ Concrete Slab: Portland Cement (sacks)	=	6	12	18	24
Sand (pounds)	=	1213	2426	3639	4852
Gravel (pounds)	=	1765	3530	5295	7060
Base gravel (pounds)	=	3640	7280	10,920	14,560

(1) Other lengths, such as 10, 12 and 14 ft. can be used.

327

Table 17-3. Brick Quantity and Cubic Feet of Mortar.

Col. 1	Patio Brick and Mortar					Wall Brick and Mortar			
	Brick Units		Mortar (cu. ft.)			Brick Units		Mortar (cu. ft.)	
Area of Patio or Wall (sq. ft.)	Col. 2	Col. 3	Col. 4	Col. 5		Col. 6	Col. 7	Col. 8	Col. 9
	4" × 8" solid pavers	3¾" × 8" solid standards	½" joints with 3¾ × 8 units	½" mortar leveling bed		8" walls	4" walls	8" walls	4" walls
1	4.5	4.0	.033	.042		12.32	6.16	.186	.072
2	9.0	8.0	.066	.083		24.64	12.32	.371	.144
3	13.5	12.0	.099	.125		36.96	18.48	.557	.216
4	18.0	16.0	.132	.167		49.28	24.64	.743	.288
5	22.5	20.0	.165	.209		61.60	30.80	.929	.360
10	45.0	40.0	.330	.417		123.20	61.60	1.857	.720
25	112.5	100.0	.825	1.043		308.00	154.00	4.643	1.800
50	225.0	200.0	1.650	2.085		616.00	308.00	9.285	3.600
75	337.5	300.0	2.475	3.128		924.00	462.00	13.928	5.400
100	450.00	400.0	3.300	4.170		1232.00	616.00	18.570	7.200

to obtain proper curing. Keep the concrete damp for at least a week before you lay the brick.

Remove the forms and position the 2 × 4 redwood frame. Remember that the frame should extend 2¾ inches above the slab. Drive redwood stakes into the ground at approximately 4-foot intervals and nail the frame to the stakes.

The techniques for laying mortared paving (Fig. 17-15) are quite similar to the methods used for building the planter. Mix 1 part portland cement, one-quarter part hydrated lime and 3 parts sand for mortar. Begin paving one corner by laying a half inch bed of mortar that is large enough for two bricks. Butter the brick sides and edges and position them according to the pattern you prefer. Continue the pattern with a half inch mortar joint between each brick.

Joints can be tooled or they can be left flush with the bricks. Use a damp cloth to wipe excess mortar from the bricks as you work. After completion, keep the patio damp for several days to obtain proper curing.

Tools

Shovel
Wheelbarrow
Garden hose
Garden rake
Framing square
2-foot hand level

Fig. 17-14. A wooden float.

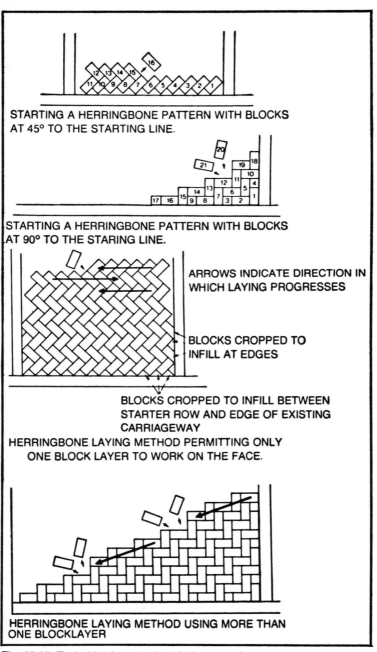

STARTING A HERRINGBONE PATTERN WITH BLOCKS AT 45° TO THE STARTING LINE.

STARTING A HERRINGBONE PATTERN WITH BLOCKS AT 90° TO THE STARING LINE.

ARROWS INDICATE DIRECTION IN WHICH LAYING PROGRESSES

BLOCKS CROPPED TO INFILL AT EDGES

BLOCKS CROPPED TO INFILL BETWEEN STARTER ROW AND EDGE OF EXISTING CARRIAGEWAY

HERRINGBONE LAYING METHOD PERMITTING ONLY ONE BLOCK LAYER TO WORK ON THE FACE.

HERRINGBONE LAYING METHOD USING MORE THAN ONE BLOCKLAYER

Fig. 17-15. Typical brick paving installation procedures.

Mason's string
Mason's line level
Hammer
Bricklayer's trowel
Wooden float (long handle)
Hand wooden float
Mason's pointing tool (or short section of ¾-inch steel pipe)
Broom (stiff bristled)
Broad-bladed brick chisel
Tamper
Screed board
2 × 4 straightedge
Burlap or straw
Powered mortar mixer (optional)

Materials

To estimate the amount of material needed to construct the patio and planter, refer to Tables 17-1, 17-2 and 17-3. If the climate in your area includes freeze/thaw cycles, use *SW* grade brick. Severe weather brick is capable of resisting harsh climatic changes.

Be sure to plan each step ahead of time. Study each diagram and all the instruction in this chapter before you actually begin constuction.

Glossary

Glossary

abutment. A masonry mass supporting the pressure of an arch, vault, beam or strut.

adhesive strength. The quality of the bond that mortar has for holding two masonry units together.

agglomerated mass. A mixture of aggregate with plaster, mortar and concrete.

aggregate. Various hard, inert material such as sand, gravel or pebbles in various size fragments mixed with cementing material to form concrete, mortar or plaster.

alignment. A parallel or converging line of upright masonry units.

amalgamation. A mixed blend or combination of materials such as lime, cement, sand and water.

American bond. A masonry bond with a course of headers between five or six courses of stretchers. Also called common bond.

anchor. A metal tie or strap that binds one part of a structure to another.

angle iron. Iron or steel bar in the form of an angle used to hold lintel units.

anta. A rectangular pier or pilaster formed by thickening the end of a masonry wall.

arcade. A series of arches supported on piers or columns.

arch. A curved structural support spanning an opening and resolving vertical load pressure into horizontal thrust.

arch buttress. See flying buttress.

arris. The sharp edge or salient angle formed by the meeting of two plane or curved surfaces.

ashlar masonry. Cut, sawed, tooled or dressed stone used for facing a wall or rubble or brick.

ashlar line. The outer line of an exterior wall above any projecting base.

axis. In masonry arch work, the center or point from which a circle or arc is formed and drawn on a template.

backfill. To replace earth around a foundation wall.

backing. Unsquared stone, rubble, brick, hollow tile or concrete block used as masonry wall to support a structure.

balanced. A proportionate distribution of the load.

basement. The substructure of a building that is wholly or partially below ground level.

bat. A portion of a brick that is whole at one end and with the other end broken off.

batter. A receding slope to a wall.

batter boards. One of a pair of boards positioned outside the corner of an excavation to indicate the proper level of the structure.

batter stick. A tapered instrument used with a level to build a battered wall.

bearing stone. Stone that withstands weight, thrust or strain.

bearing wall. A partition which supports the weight of a structure.

bed. The surface on which brick is laid.

bed joint. A horizontal masonry joint.

bed stone. A large foundation stone.

belt course. A horizontal band around pillars or columns.

bench mark. A fixed elevation mark or point of reference on a concrete post set in the ground from which measurements can be made.

binder course. A course of aggregate between the foundation and the pavement.

binders. A substance such as cement that promotes cohesion of materials.

blind bond. A masonry bond in which headers extend only half of the way through the tier of the face brick. The face bricks are all stretchers and some are split lengthwise to accomodate the headers.

block. A solid unit of masonry material formed in a uniform size.

block and bond. Laying a block bond on one side and a different unit and bond on the other side.

blocking course. The finished course of a wall laid on top of a cornice to give weight and to bind the cornice.

boasting. Cutting masonry with a broad-edged chisel.

bond. The systematic lapping of brick or other masonry units to enhance strength and appearance.

bond course. A course of masonry bondstone laid crossways partially or entirely through a wall.

boning rods. Rods to sight across masonry to keep the units level as they are laid on a horizontal surface.

breadth. The width of a masonry wall.

breaking load. The stress or tension sufficient to break or rupture.

breast (fireplace). The front of a fireplace.

brick. Building or paving material made by baking or burning molded clay into blocks.

brick beam. A lintel of bricks with iron straps.

brick chisel. A broad-edged chisel used to trim or cut brick.

brick set. Another name for a brick chisel.

brick trowel. A flat triangular instrument used in bricklaying to spread mortar or concrete.

brick veneer. Brick facing bonded to a wall built of another material.

bricklayer. A person who constructs buildings, chimneys or other structures made of brick or block and mortar.

bricklayer's hammer. A hammer with a flat face and sharp peen used to dress or break brick.

brickwork. A structure built by laying brick with or without mortar.

bridge stone. Masonry that spans a gutter.

bridge wall. A low, separating wall of firebrick.

broken range. Random laying of masonry units.

brooming. Applying a finish to concrete with a broom.

buck. A rough doorframe placed in a wall or partition during construction.

buckstay. Either of two connected girders, one on each side of a masonry structure to take the thrust of an arch.

bulging. The spreading of a wall usually caused by excessive load pressure.

bull header. A header brick laid on edge with the end exposed.

bull stretcher. Brick laid on edge with the face exposed.

bull's-eye. A circular or oval opening in a wall.

bush hammer. A hammer with a serated face for dressing masonry units and concrete.

burning. Processing brick in a kiln at 100 to 400 degrees.

buttered joint. A thin layer of mortar applied to one end of a brick before it is laid.

buttress. A projecting structure of masonry that resists lateral pressure at a specific point.

clacine. To heat materials for manufacturing cement or lime.

camber. A slight convex arch or curve.

carve. To cut or dress brick or stone.

casting. Molded or shaped masonry.

cavity wall. A masonry wall built in two thicknesses separated by an air space designed to provide thermal insulation.

cell. An air space in a hollow tile or a cement block.

cement. A powdered mixture of alumina, silica, lime, iron oxide and magnesia burned together in a kiln and finely pulverized. When it is mixed with water it forms a plastic mass that hardens by chemical combination.

chain bond. A bond formed by building in a tie or strap.

chain course. A course of continuously fastened headers held by cramps.

chase. A groove or shallow channel in masonry for a pipe or conduit.

chimney. A vertical masonry structure for carrying off smoke.

chimney breast. The portion of a chimney or fireplace that projects from a wall into a room.

chimney cap. A device used to improve the draft from a chimney.

chimney hood. A protective covering over the indoor portion of a chimney.

clastic. Conglomerate stone.

cleavage plane. The natural division of masonry units that results in a sharp division or splitting in specific directions.

clinker brick. The result when brick is over burned in a kiln.

clip bond. A masonry wall bond formed by clipping off the inner corners of the block or face brick to obtain a diagonal bond with stretchers.

closer. The last masonry unit in a course.

cohesive strength. The resistance to separation of masonry.

common bond. See American bond.

concrete. A solid, durable building material formed by mixing cement, aggregate and water.

concrete finishing. Applying a smooth surface to concrete.

concrete masonry. Brick, block or tile construction.

conglomerate. Rounded fragments of sedimentary rock cemented into a mass.

coping. The highest course or the covering course of a masonry wall. It is usually sloped to carry off water.

cored brick. Brick with three holes in a single row near the center or two rows of five holes.

corbel. A masonry unit that projects from a wall upward and outward.

cornice. The top course of a wall that projects horizontally.

course. One layer or row of masonry units.

course bed. The top of the last masonry course laid.

cramp. A dovetailed form used to bond concrete block.

crown. The finishing element of a masonry surface.

curing. Chemical change in concrete obtained by maintaining proper moisture and temperature conditions.

curtain wall. A nonbearing wall between piers. Also called an enclosure wall.

diagonal bond. A masonry bond in which headers are bonded to the wall by concealed parallel bricks.

dipping. Putting masonry units into mortar rather than spreading the mortar on them.

dowels. A piece of wood driven into a hole that has been drilled into masonry.

draft. A device for regulating air currents through a chimney or fireplace.

dressing. Working the face of a brick or stone.

dry stone. Stone laid without mortar.

Dutch bond. Alternate courses of headers and stretchers. Joints between the stretchers are over the centers of the stretchers.

efflorescence. The formation of a powder or incrustation on a masonry surface due to capillary transfer of soluble ground salt.

English bond. Masonry bond where header courses are alternated with stretcher courses.

face brick. Brick commonly used for exposed surfaces. Manufactured in various colors.

fire brick. Brick made from fireclay that can withstand higher temperatures than regular brick.

fireclay. A refractory clay used in making firebrick.

fire wall. Masonry placed between joists to prevent the spread of fire.

Flemish bond. A masonry bond in which each course alternates headers and stretchers so that each course is centered above and below a stretcher.

flue. A passage or duct for chimney smoke.

flush. To bring a masonry unit even with the surface of the structure.

flying buttress. An arch spanning a passageway to a solid pier or buttress.

footing. Foundation or bottom unit of a wall.

form. A temporary frame that retains concrete until the concrete hardens.

forming. Tempering clay to produce a homogeneous mass for making brick.

foundation. The portion of a structure, usually below ground level, that supports a wall or other structure.

galleting. Filling a masonry joint with rock chips to increase the strength of the joint.

garden wall bond. Alternating two or more stretchers in each course.

graded aggregate. Aggregate with various types of sand and gravel.

grout. Mortar that has been thinned with water so that it can be poured.

header. Masonry unit laid with the shorter ends exposed.

header bond. Masonry bond in which all courses are overlapping headers.

header course. A row of headers.

hearth. The floor of a fireplace, usually made of stone or brick.

herringbone bond. Masonry bond in which the exposed bricks are laid diagonally to the wall.

interlocking. Bonding masonry units by lapping them.

joggle. To fit or fasten with dowels.

joint. The space between two adjacent masonry units bonded by mortar.

jointer. A tool used for making joints.

keystone. Masonry unit at the summit of an arch.

kneeler. A brick or stone supporting inclined masonry.

lateral thrust. Movement caused by load pressure.

lean mortar. Mortar that is deficient in bonding material.

lime. When combined with water, lime forms calcium carbonate. Used in mortar and cement.

lintel. Horizontal architectural member supporting the weight above a door or window.

mortar. Mixture of lime, cement and water used to bond masonry.

mortar board. A small, lightweight board on which masons temporarily place mortar.

muriatic acid. Hydrochloric acid diluted with water and used to clean masonry.

neat cement. Cement and water without aggregate.

niche. A recess in a wall.

nogging. To fill open spaces of a wood frame with bricks or other masonry.

offset. To build a wall with a reduction in thickness.

parapet. An elevation raised above the main wall, generally at the edge for decoration.

pargeting. A lining of mortar or plaster for a chimney.

parging. A thin, smooth coat of mortar applied to masonry.

pier. A portion of a wall between windows or doors.

pilaster. A shallow rectangular projection from a wall, usually in the shape of a column.

pitch. To square a masonry surface by cutting with a chisel.

plumb. Having a perpendicular or vertical line that is true.

pointing. Filling masonry joints with mortar. Dress the surface of masonry with a tool.

portland cement. Hydraulic cement made by burning a mixture of limestone and clay in a kiln.

precast. To cast a concrete block or slab before placing it in a structure.

pugging. To knead clay with water to make it plastic. Fill with clay to deaden sound.

putlog. A short horizontal support.

racking bond. Stepping back the ends of a course from bottom to top on an unfinished wall.

rich mix. Mortar with an excess of bonding material.

riprap. Broken stone for foundations. A masonry wall with an irregular pattern.

rowlock. A brick laid on edge as a header.

rubble. Broken stone of irregular size and shape.

sand. Small, loose grains (usually quartz) used as aggregate.

sand cushion. A layer of sand separating paving and subsoil.

sandblasting. A blast of air or steam with sand used to clean masonry.

scaffold. A temporary structure for holding workers and material during construction.

screed. A wooden strip for smoothing the surface of concrete.

shoved joint. A joint made by shoving a masonry unit into position.

slump test. The method for determining the consistency of concrete.

soldier course. A masonry course in which the units are laid vertically with the narrow face exposed.

split. A brick half the normal thickness and used to support a course of bricks over another course that is not level.

stack bond. Masonry units positioned one on top of the other and having the same vertical joints.

story pole. A rod cut equal to the height of one story of a building with markings indicating specific heights.

straightedge. A strip of wood or metal used to test for straight lines.

stretcher. A masonry unit with its length parallel to the face of the wall.

stretcher bond. See running bond.

string course. A course of stretcher masonry units.

tempering. To bring to a proper consistency by mixing or blending.

template. A pattern, usually of wood or metal, used to make a copy of an object.

tensile strength. The resistance of a material to stress.

texture. The visual and tactile quality of a masonry surface.

tooled. Worked or shaped masonry.

tucking. Filling mortar joints after masonry units are laid.

tuck-pointing. Pointing which has an ornamental line projecting from the structure.

winning. Mining procedure where several days production are held in storage.

wythe. A continuous vertical tier of masonry. The partition dividing flues of a chimney.

Appendices

Directory of Brick Manufacturers

Abilene Brick Co.
Box 1107
Abilene, TX 79604

Acme Brick Co. (main office)
P.O. Box 425
Fort Worth, TX 76101
 Other offices or plants located in Fort Smith, Arkansas; Malvern, Arkansas (2 plants); Perla, Arkansas; Kanopolis, Kansas, Weir, Kansas; Baton Rouge, Lousiana; Clinton, Oklahoma; Edmond, Oklahoma (2 plants); Tulsa, Oklahoma; Bridgeport, Texas; Denton, Texas; Garrison, Texas; McQueeney, Texas; Millsap, Texas.

Ada Brick Co.
Box 501
Ada, OK 74820

Alamo Clay Products Co.
P.O. Box 93
Elmendorf, TX 78112

Alton Brick Co.
2510 Adie Road
St. Louis, MO 63043

American Brick Co.
6558 W. Fullerton Ave.
Chicago, IL 60635
 Also in Riverdale, Illinois and Munster, Indiana

Andy Cordell Brick Co.
7226 North Loop
Houston, TX 77028

Arizona Brick Co., Inc.
P.O. Box 17946
Tucson, AZ 85731

Ashe Brick Co.
Van Wyck, SC 29744

Athens Brick Co.
P.O. Box 70
Athens, TX 75751

Atkinson Brick Co.
13633 S. Central Ave.
Los Angeles, CA 90059

Baltimore Brick Co.
110 West Road
Baltimore, MD 21204
    Also Rocky Ridge, Maryland and Rossville, Maryland

Barboursville Clay Manufacturing Co.
P.O. Box 691
Nitro, WV 25143

The Belden Brick Co. (main office)
700 Tuscarawas St. W.
Canton, OH 44701
    Also Port Washington, Ohio; Strasburg, Ohio; Sugercreek, Ohio (5 plants); Uhrichsville, OH

Bennett Brick & Tile Co.
P.O. Box 29
Kings Mountain, NC 28086

Bickerstaff Clay Products Co., Inc.
1338 Fourth Ave.
Columbus, GA 31902
    Also at Bessemer, Alabama; Phenix City, Albama (2 plants); Smyrna, Georgia.

Binghamton Brick Co., Inc.
P.O. Box 1256
Binghamton, NY 13902

Birmingham Clay Products
#2 Office Park Circle, Suite 101
Birmingham, AL 35223

Bloomfield Shale Inc.
P.O. Box 272
Bloomfield, IN 47424

Borden Brick & Tile Co.
P.O. Box 11558
Durham, NC 27703
    Also at Sanford, North Carolina.

Boren Clay Products Co.
P.O. Box 368
Pleasant Garden, NC 27313
    Also Charlotte, North Carolina; Monroe, North Carolina; Roseboro, North Carolina; Gaffney, South Carolina.

The Bowerston Shale Co.
Box 199
Bowerston, OH 44695
    Also at Newark, Ohio.

Brick & Tile Corp. of Lawrenceville
Box 45
Lawrenceville, VA 23868

Broad River Brick Co.
P.O. Box 550
Gaffney, SC 29340
    Also at Blacksburg, South Carolina.

Brookhaven Pressed Brick Co.
P.O. Box 1402
Brookhaven, MS 39601

Burley Brick and Sand Co.
1010 E. Main
Burley, ID 83318

Burns Brick Co.
711 10th St.
Macon, GA 31208

Burns Clay Products
P.O. Box 61
Newton, AL 36352

Can-Tex Industries
Masonry Products Division
Box 469
W. Des Moines, IA 50265
    Also at Grimes, Iowa; Ottumwa, Iowa; Redfield, Iowa.

Carolina Ceramics Inc.
RFD 3, Box 266
Columbia, SC 29206

Castaic Clay Manufacturing Co.
P.O. Box 8
Castaic, CA 91310

Chatham Tile & Brick Co.
P.O. Box 87
Gulf, NC 27256

Cheraw Brick Works Inc.
P.O. Box 207
Cheraw, SC 29520

Cherokee Brick Company of North Carolina
P.O. Box 33218
Raleigh, NC 27606
    Also at Brickhaven, North Carolina.

Cherokee Brick & Tile Co.
Box 4567
Macon, GA 31208

Chestertown Brick Co.
P.O. Box 28
Chestertown, MD 21620

Clay Products Inc.
P.O. Box 597
Holly Springs, MS 38635

Clayburn Inc.
P.O. Box 2828
Clairmont Station
Everett, WA 98203

Claycraft Co.
698 Morrison Road
Columbus, OH 43216
    Also at Sugercreek, Ohio and Upper Sandusky, Ohio.

Cleveland Builders Supply Co.
2100 West 3rd
Cleveland, OH 44113

Cline Brick Co.
P.O. Box 1790
Ashland, KY 41101

Cloud Ceramics
Box 369
Concordia, KS 66901

Colonial Brick Corp.
Box 365
Cayuga, IN 47928

Colorado Brick Co.
P.O. Box 3513
Boulder, CO 80303

Columbia Brick & Tile Co.
Box 246
Columbia, MO 65201

Columbus Brick Co.
P.O. Box 866
Columbus, MS 39701

Commercial Brick Corp.
P.O. Box 1382
Wewoka, OK 74884

Continental Clay Products Co.
Box 69
Kittanning, PA 16201

Corbin Brick Co., Inc.
Box 452
Corbin, KY 40701
    Also at Woodbine, Kentucky.

Craycroft Brick Co.
2301 W. Belmont
Fresno, CA 93728

Cunningham Brick Co., Inc.
Thomasville, NC 27360

Victor Cushwa & Sons
P.O. Box 228
Williamsport, MD 21795

Dalton Bros. Brick Co., Inc.
N. Elm St. & Means Ave.
Hopkinsville, KY 42240

Darlington Brick & Clay Products Co.
1910 Cochran Road
Manor Oak Building 1
Pittsburgh, PA 15220

Davidson Brick Co.
2601 W. Floral Drive
Monterey Park, CA 15220

Delaware Brick Co.
1220 Centerville Road
Wilimington, DE 19808
    Also at Dover, Delaware and New Castle, Delaware

De Leon Brick Inc.
P.O. Box 187
De Leon, TX 76444

Delta Brick & Tile Co., Inc.
P.O. Box 539
Indianola, MS 38751

Delta-Macon Brick & Tile Co., Inc.
Rt. 4, Box 2
Macon, MS 39341

Delta-Shuqualak Brick & Tile Co., Inc.
P.O. Box 67
Shuqualak, MS 39361

Dennis Brick Co.
9 Brickyard Circle
Auburn, ME 04210

The Denver Brick & Pipe Co.
Box 2329
Denver, CO 80201

D'Hanis Brick & Tile Co.
P.O. Box 397
D'Hanis, TX 78850

Dixie Brick Co.
Box 969
Natchitoches, LA 71457
    Also at Jamestown, Louisiana.

Dorchester Brickworks
P.O. Box 755
Summerville, SC 29483

Elgin-Bulter Brick Co.
P.O. Box 1947
Austin, TX 78767
    Also two plants at Butler, Texas.

El Paso Brick Co., Inc.
P.O. Box 12336
El Paso, TX 79912

Endicott Clay Products Co.
P.O. Box 17
Fairbury, NE 68352
    Also at Endicott, Nebraska.

Entrada Industries
Interstate Brick Division
P.O. Box 517
West Jordan, UT 84084

Eureka Brick & Tile Co.
Box 379
Clarksville, AR 72830

Evans Clay Products Inc.
Barnhill Road
Midvale, OH 44653

Excelsior Brick
P.O. Box 32
Fredonia, KS 66736

Excelsior Brick Co., Inc.
1228 N. McDonough St.
Montgomery, AL 36104

Fairfield Brick Co.
Box 115
Zoarville, OH 44698

Fairhope Clay Products Inc.
Rt. 1, Box 78
Fairhope, AL 36532

Florida Brick & Clay Co., Inc.
P.O. Box 1656
Plant City, FL 33566

Frame Brick & Tile Co.
P.O. Box 1863
Anniston, AL 36201

Freeport Brick Co.
273 Bailles Run Road
Allegheny Brick Division
Creighton, PA 15030

The Galena Shale Tile & Brick Co.
P.O. Box 188
Galena, OH 43021

General Clay Products Corp.
1445 W. Goodale Blvd.
Columbus, OH 43212
Also at Baltic, Ohio; Logan, Ohio; Nelsonville, Ohio; New-comerstown, Ohio.

General Shale Products Corp.
Box 3547, C.R.S.
Johnson City, TN 37601
Also at Huntsville, Alabama; Atlanta, Georgia; Mooresville, Indiana; Fairdale, Kentucky; Chattanooga, Tennessee; Kingsport, Tennessee; Knoxville, Tennessee; Glasgow, Virginia; Marion, Virginia; Richlands, Virginia.

General Wadsworth Brick Corp.
1445 W. Goodale Blvd.
Columbus, OH 43212

Georgia-Carolina Brick & Tile Co.
P.O. Box 1957
Augusta, GA 30903

Glen-Gery Corp.
277 N. 5th St.
Reading, PA 19603
Also at New Oxford, Pennsylvania; Shoemakersville, Pennsylvania; Watstontown, Pennsylvania; Wyomissing, Pennsylvania; York, Pennsylvania.

Goodwin Tile & Brick Co.
P.O. Box 283
Des Moines, IA 50341

Guignard Brick Works
P.O. Box 568
Lexington, SC 29072

Handford Brick Co., Inc.
P.O. Box 1215
Burlington, NC 27215

Hanley Co.
28 Kennedy St.
Bradford, PA 16701.
Also at Lewis Run, Pennsylvania; Summerville, Pennsylvania

Hattiesburg Brick Works Inc.
2600 Lakeview Road
Hattiesburg, MS 39401

Hazleton Brick Inc.
Box 355 Hazle Village
Hazleton, PA 18201

Hebron Brick Co.
Hebron, ND 58638

Henderson Clay Products Inc.
Box 1129
Henderson, TX 75652
    Also at Marshall, Texas.

Henry Brick Co., Inc.
P.O. Box 857
Selma, AL 36701

Herbert Materials Inc.
1136 2nd Ave. N.
Nashville, TN 37208
    Also at Gleason, Tennessee.

Hidden Brick Co.
2610 Kauffman Ave.
Vancouver, WA 98660

Higgins Brick Co.
P.O. Box 7000-167
Redondo Beach, CA 90277

Hillabee Brick Co., Inc.
P.O. Box 656
Alexander City, AL 35010.

Holly Springs Brick & Tile Co., Inc.
P.O. Box 310
Holly Springs, MS 38635

Hope Brick Works
P.O. Box 663
Hope, AR 71801

Houston Brick & Tile Co.
6614 John Ralston
Houston, TX 77049

Humboldt Brick & Tile Co.
Box 185
Humboldt, KS 66748

Interpace Corp.
401 2nd Ave. W.
Seattle, WA 98119
    Also at Ogden, Utah; Mica, Washington; Renton, Washington.
Isenhour Brick & Tile Co.
P.O. Box 1249
Salisbury, NC 28144
    Also at East Spencer, North Carolina.

Jenkins Brick Co., Inc.
1605 Furnace, Box 91
Montogomry, AL 36101
    Also at Coosada, Alabama; Atlanta, Georgia.

K-F Brick Co., Inc.
E. Windsor Hill, CT 06028
    Also at Middleboro, Massachusetts.
Kane-Gonic Brick Corp.
Winter St.
Gonic, NH 03867
Kansas Brick & Tile Co., Inc.
P.O. Box 414
Hoisington, KS 67544
Kasten Clay Products Inc.
Box 347
Jackson, MO 63755
Keego Clay Products Co., Inc.
Rt. 2
Brewton, AL 36426
Kentucky Brick Co.
P.O. Box 741
Owensboro, KY 42301
Kentwood Brick Co.
P.O. Drawer F
Kentwood, LA 70444
Kings Mountain Brick Inc.
P.O. Box 683
Kings Mountain, NC 28086
Kinney Brick Co., Inc.
P.O. Box 1804
Albuquerque, NM 87103
Klamath Falls Brick & Tile Co.
Box 573
Klamath Falls, OR 97601

Lachance Brick Co.
Gorham, ME 04038
Laird Brick Co., Inc.
P.O. Box 98
Puryear, TN 38251
The Lakewood Brick & Tile Co.
1325 Jay St.
Lakewood, Co. 80214
Leesburg Brick Co., Inc.
Rt. 1, Box 83A
Leesburg, TX 75451
Laredo Brick Co., Inc.
Box 381
Laredo, TX 78040
Laurel Brick & Tile Co., Inc.
Box 583
Laurel, MS 39440
Louisville Brick Co.
Hwy. 15N Box 426
Louisville, MS 39339
The Lovell Clay Products Co.
P.O. Box 2096
Billings, MT 59103
Lee Brick & Tile Co., Inc.
P.O. Box 1027
Sanford, NC 27330
The McAvoy Vitrified Brick Co.
Phoenixville, PA 19460
L.P. McNear Brick Co., Inc.
P.O. Box 1380
NcNear Point
San Rafael, CA 94902
The Michael Kane Brick Co., Inc.
654 Newfield St.
Middletown, Co. 06457
McNees-Kittanning Co.
260 Oak Ave.
Kittanning, PA 16201
Malvern Flue Lining Inc.
Box 465
Malvern, OH 44644

Mangum Brick Co.
P.O. Box 296
Mangum, OK 73554

Marion Brick Corp.
P.O. Box 296
Marion, OH 43302
  Also at Brazil, Indiana; Caledonia, Ohio; Morral, Ohio; Bigler, Pennsylvania; Clearfield, Pennsylvania; Manassas, Virginia.

Martin Brick Co.
Box 980
Coleman, TX 86834

Maryland Clay Products Inc.
7100 Muirkirk Road
Beltsville, MD 20705

Medora Brick Co., Inc.
N. Ewing St.
Brownstown, IN 47220
  Also at Medora, Indiana.

Merry Companies Inc.
P.O. Box 1957
August, GA 30903
  Also at Columbia, South Carolina; Summerville, South Carolina.

Michigan Brick Inc.
P.O. Box 66
Corunna, MI 48817
  Also at Stanton, Kentucky; Mansfield, Ohio; Mineral Wells, Texas.

Midland Brick & Tile Co.
Box 428
Chillicothe, MO 64601

Mike-Baker Brick Co., Inc.
P.O. Box 51234
Lafayette, LA
  Also at Cade, Louisiana.

Milliken Brick Co.
Rich Hill Road
Cheswick, PA 15024
  Also at Darlington, Pennsylvania; Harmar Township, Pennsylvania; Pitcairn, Pennsylvania.

Mineral Wells Clay Products
P.O. Box 369
Mineral Wells, TX 76067

Mission Valley Brick Co.
Box 3217
San Diego, CA 92103

The Moland-Drysdale Corp.
Box 2150
Henderson, NC 28739
    Also at Fletcher, North Carolina.

Monroe Clay Products
P.O. Box 78
Monroe, OR 97456

Morin Brick Co.
Danville, ME 04223

H.C. Muddox Co.
4875 Bradshaw Road
Sacramento, CA 85826
    Also at Ione, California.

Mutual Materials Co.
P.O. Box 209
Bellevue, WA 98009

Nash Brick Co., Inc.
Box 962
Rocky Mount, NC 27801
    Also at Ita, Halifax County, North Carolina.

Nassau Brick Co., Inc.
635 Round Swamp Road
Bethpage, NY 11804

New Jersey Shale Brick Corp.
Hamilton Road
Somerville, NJ 08870

Oakfield Shale Brick & Tile Co.
P.O. Box 337
Oakfield, WI 53065

Ochs Brick & Tile Co.
P.O. Box 106
Springfield, MN 56087

Oklahoma Brick Corp.
4300 N.W. Tenth St.
Oklahoma City, OK 73147
     Also at Union City, OK.

Old Carolina Brick Co.
Rt. 9, Box 77
Salisbury, NC 28144

Old Virginia Brick Co., Inc.
Box 508
Salem, VA 24153

Omaha Brick Works Inc.
72nd & Main
Ralston, NE 68127

Otto Brick & Tile Works Inc.
Main Road
Springs, PA 15562

Owensboro Brick & Tile
P.O. Box 708
Owensboro, KY 42301

Pacific Clay Building Products
9500 South Norwalk Blvd.
Santa Fe Springs, CA 90670

Palmetto Brick Co.
Box 430
Cheraw, SC 29520

Payne Brick Co.
P.O. Box 189
Elgin, TX 78621

Pee Dee Ceramics Inc.
Drawer B
Pee Dee, SC 29586

Peoria Brick & Tile Co.
Box 515
Peoria, IL 61651

Phoenix Brick Yard
P.O. Box 768
Phoenix, AZ 85007

Pine Hall Brick & Pipe Co.
P.O. Box 11044
Winston-Salem, NC 27106
    Also at Madison, North Carolina.

Port Costa Products Co.
P.O. Box 5
Port Costa, CA 94569

Powell & Minnock Brick Works Inc.
Wolf Road Park
Metro Park Road
Albany, NY 12205

Pullman Brick Co.
5657 Warm Springs Ave.
Boise, ID 83706

Quakertown Brick & Tile Co., Inc.
Heller Road
Quakertown, PA 18951

Ragland Brick Co., Inc.
3507 Rainbow Road
Gadsden, AL 35901

Redford Brick Co., Inc.
Box 4096
Richmond. VA 23224

Reliance Clay Products
2947 Blystone
Dallas, TX 75220
    Also at Ferris, Texas; Lindale, Texas.

Richards Brick Co.
Box 407
Edwardsville, IL 62025

Richland Brick Co.
P.O. Box 457
Mansfield, OH 44901

Richtex Corp.
Box 3307
Columbia, SC 29230
    Also at Sumter, South Carolina.

Robinson Brick & Tile Co.
P.O. Box 5243
Denver, CO 80217

Royal River Brick Co., Inc.
Box 458
Gray, ME 04039
    Also at North Yarmouth, Maine.

St. Joe Brick Works Inc.
P.O. Box 400
Slidell, LA 70459

Salisbury Brick Corp.
P.O. Drawer S
Summerville, SC 29483

Sanford Brick Corp.
P.O. Drawer 458
Sanford, NC 27330
    Also at Gulf, North Carolina; Norwood, North Carolina.

San Luis Brick Inc.
2900 S. Broad St.
San Luis Obispo, CA 93401

Sapulpa Brick & Tile Corp.
704 W. Dewey
Sapulpa, OK 74066

Savery Brick & Tile Co.
P.O. Box 560
Tupelo, MS 38801
    Also at Baldwyn, Mississippi.

Sheffield Brick & Tile Co.
P.O. Box 676
Sheffield, IA 50475

Sioux City Brick & Tile
222 Commerce Building
Sioux City, IA 51102
    Also at Sergeant Bluff, Iowa.

Sipple Brick Inc.
P.O. Box 567
Stanton, KY 40380

Southern Brick Co., Inc.
Byhalia, MS 38611

Southwestern Brick Inc.
Rt. 2, Box 167
Snyder, TX 79549

Stanley Shale Products
P.O. Box 505
Norwood, NC 28128

Stark Ceramics Inc.
P.O. Box 8880
Canton, OH 44711

Statesville Brick Co.
P.O. Box 471
Statesville, NC 28677

Stiles & Hart Brick Co.
Box 367
Bridgewater, MA 02324

Stone Creek Brick Co.
Box 116
Stone Creek, OH 43840

Streator Brick Systems Inc.
P.O. Box E
Streator, IL 61364

Structural Stoneware Inc.
P.O. Box 119
Minerva, OH 44657

Summit Pressed Brick & Tile Co.
P.O. Box 533
Pueblo, CO 81002

Taylor Clay Products Co.
P.O. Box 2128
Salisbury, NC 28144

Teague Brick Sales Co.
P.O. Box 329
Teague, TX 75860

Texas Clay Industries
P.O. Box 469
Malakoff, TX 75148

The Texas Brick Co., Inc.
P.O. Box 398
Brownwood, TX 76801

Triangle Brick Co.
Box 60
Triangle Park Area, Rt. 4,
Durham, NC 27713

Tri-State Brick & Tile Co., Inc.
P.O. Box 9787
Jackson, MS 39206

United Brick & Tile Co.
Adel, IA 50003

Waccamaw Clay Products Co., Inc.
P.O. Box 2679
Myrtle Beach, SC 29577

Watkins Brick Co., Inc.
P.O. Box B
Birmingham, AL 35218
    Also at Ensley, Alabama.

Watsontown Brick Co.
R.D. 2, Box 68
Watsontown, PA 17777

Webster Brick Co., Inc.
Box 12887
Roanoke, VA 24029
    Also at Eden, North Carolina; Somerset, Virginia; Suffolk,
Virginia; Webster, Virginia.

Wheeler Brick Co., Inc.
P.O. Box 162
Jonesboro, AR 72401

Whitacre-Greer Fireproofing Co.
E. Lisbon St.
Waynesburg, OH 44688
    Also at Alliance, Ohio; Magnolia, Ohio.

Yadkin Brick Co., Inc.
New London, NC 28127

Yankee Hill Brick Manufacturing Co.
3705 S. Coddington Ave.
Lincoln, NE 68502

Useful Charts

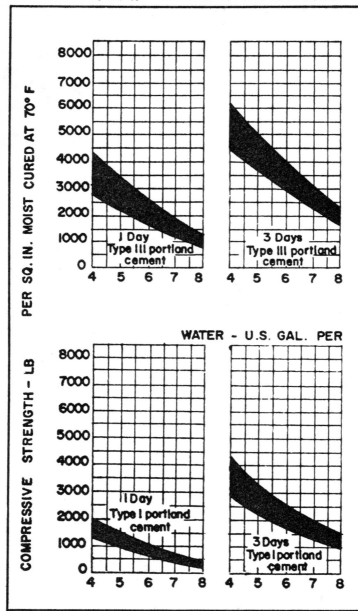

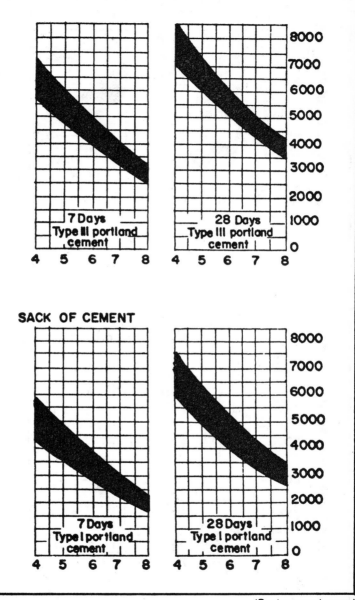

SACK OF CEMENT

(Cont. on next page.)

(Cont. from previous page.)

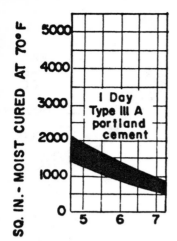

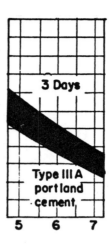

WATER U.S. GAL. PER SACK

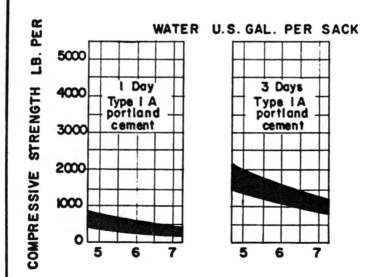

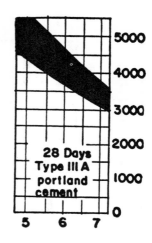

OF CEMENT

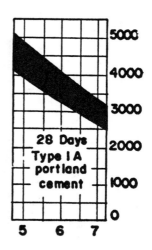

Table B-2. Approximate Mixing Requirements for Slumps and Maximum Sizes of Aggregates.

Maximum size of aggregate, in.	Air-entrained concrete				Non-air-entrained concrete			
	Recommended average total air content, per cent†	Slump, in.			Approximate amount of entrapped air, per cent	Slump, in.		
		1 to 2	3 to 4	5 to 6		1 to 2	3 to 4	5 to 6
		Water, gal. per cu. yd. of concrete**				Water, gal. per cu.yd. of concrete**		
⅜	7.5	37	41	43	3.0	42	46	49
½	7.5	36	39	41	2.5	40	44	46
¾	6.0	33	36	38	2.0	37	41	43
1	6.0	31	34	36	1.5	36	39	41
1½	5.0	29	32	34	1.0	33	36	38
2	5.0	27	30	32	0.5	31	34	36
3	4.0	25	28	30	0.3	29	32	34
6	3.0	22	24	26	0.2	25	28	30

†Adapted from Recommended Practice for Selecting Proportions for Concrete (ACI 613-54).
**These quantities of mixing water are for use in computing cement factors for trial batches. They are maximums for reasonably well-shaped angular coarse aggregates graded within limits of accepted specifications.
†Plus or minus 1 per cent.

Table B-3. Wind Pressures for Various Height Zones.

Height Zone, ft	Wind-Pressure-Map Areas, psf						
	20	25	30	35	40	45	50
Less than 30	15	20	25	25	30	35	40
30 to 49	20	25	30	35	40	45	50
50 to 99	25	30	40	45	50	55	60
100 to 499	30	40	45	55	60	70	75
500 to 1199	35	45	55	60	70	80	90
1200 and over	40	50	60	70	80	90	100

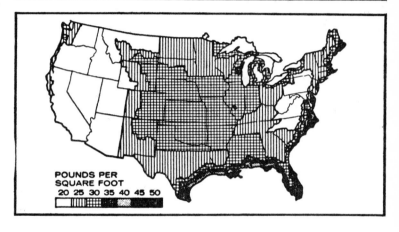

POUNDS PER
SQUARE FOOT
20 25 30 35 40 45 50

Index

Index